AP* SUCCESS

PHYSICS B/C

2001

Boost Your Score on the AP Exams in Physics B/C

John W. Dooley, Ph.D.
Matthew G. Sexton, M.S.
Steven O. Nelson, M.S.
Gabriel Lombardi, Ph.D.
Jay Streib, Ph.D.

www.petersons.com

PETERSON'S
THOMSON LEARNING

Australia • Canada • Mexico • Singapore • Spain • United Kingdom • United States

PETERSON'S
THOMSON LEARNING

About Peterson's

Founded in 1966, Peterson's, a division of Thomson Learning, is the nation's largest and most respected provider of lifelong learning online resources, software, reference guides, and books. The Education Supersite℠ at petersons.com—the Web's most heavily traveled education resource—has searchable databases and interactive tools for contacting U.S.-accredited institutions and programs. CollegeQuest℠ (CollegeQuest.com) offers a complete solution for every step of the college decision-making process. GradAdvantage™ (GradAdvantage.org), developed with Educational Testing Service, is the only electronic admissions service capable of sending official graduate test score reports with a candidate's online application. Peterson's serves more than 55 million education consumers annually.

Thomson Learning is among the world's leading providers of lifelong learning, serving the needs of individuals, learning institutions, and corporations with products and services for both traditional classrooms and for online learning. For more information about the products and services offered by Thomson Learning, please visit www.thomsonlearning.com. Headquartered in Stamford, Connecticut, with offices worldwide, Thomson Learning is part of The Thomson Corporation (www.thomson.com), a leading e-information and solutions company in the business, professional, and education marketplaces. The Corporation's common shares are listed on the Toronto and London stock exchanges.

For more information, contact Peterson's, 2000 Lenox Drive, Lawrenceville, NJ 08648;
800-338-3282; or find us on the World Wide Web at: www.petersons.com/about

Editorial Development: American BookWorks Corporation

Contributing Editors: Christopher J. Cramer, Ph.D., Ponn Maheswaranatha, Ph.D.

ISSN 0894-9336
ISBN 0-7689-0459-4

Printed in the United States of America

10 9 8 7 6 5 4 3 2 1 02 01 00

CONTENTS

PRACTICE TESTS

INTRODUCTION

ABOUT THIS BOOK

The AP Physics exam may appear daunting at first, but if you've prepared for the test throughout the year and take the time to use this book properly, it should not be that difficult. We have tried to make this a workable book. In other words, the book is set up so that you will be able to find the material that is necessary to study and be fully prepared when it's time to take the actual test.

The book begins with a Physics Diagnostic Test. The purpose of the Diagnostic Test is to help you get a handle on what you know and what needs more work. We have included material from the General Physics section, as well as questions from both the Physics B and Physics C exams. Take this exam (and all of the tests) under simulated exam conditions, if you can. What this means is that you should find a quiet place in which to work, set up a clock, and take the test without stopping. When you are finished, take a break and then go back and check your answers. Always reread those questions you got wrong, since sometimes errors come from merely misreading the question. Again, double-check your answers, and if they're still not clear, read the review material. We also suggest that you answer all of the questions, regardless of the version of the exam you plan to take.

Once you've completed the Diagnostic Test, it's time to move on to the physics review. Study the material carefully, but feel free to skim portions of the review section that are easy for you. There are eighteen chapters in all. In fact, before you begin any of this work, it would be helpful to consult the suggested study plans that follow this introduction.

Then, take the actual practice tests. There are two simulated exams for Physics B and two simulated exams for Physics C. These tests are designed to give you an idea of the types of questions you will encounter on the exam. While these are not actual exams, the questions themselves are the same types of questions you will find the on the actual AP Physics tests.

As you complete each exam, take some time to review your answers. We think you'll find a marked improvement in your scores from the time you take the Diagnostic Test to the time you complete all of the full-length practice tests. As you go through the tests, circle those answers that you are not sure of, so you can go back and review them. Always take the time to check the review section for clarification, and if you still don't understand the material, go to your teacher for help.

ABOUT THE TESTS

The Physics B exam is three hours long. The first section contains 70 multiple-choice questions and the second is a free-response section that contains 6 to 8 questions. You will have 1 1/2 hours to take each section of this test.

The Physics C exam also consists of two parts, each 1 1/2 hours long. One part covers mechanics and the other part covers electricity and magnetism. You may take either part, or you may take both parts, and you will get separate grades for each section. There are 35 multiple-choice questions in each section, and each part has a free response section as well. There are usually three questions in each free response section.

TAKING THE TEST

Since you will have 3 hours in which to complete the exam, it is important to pace yourself. It is very important that you make sure you are thoroughly familiar with the directions for the tests, so that you don't have to waste time trying to understand them once you've opened your test booklet.

Work through the easy questions first. The faster you complete those questions, the more time you'll have for those that are more difficult. You may use your test book for scratch paper, but keep your answer sheet clean, since they are machine-readable, and any stray marks might be construed as an answer.

The multiple-choice questions have five lettered choices. As with any multiple-choice question, you should approach each one by first trying to select the correct answer. If the answer is clear to you, select it at once. If you're unsure, the first technique is the process of elimination. Try to cross off any answers that don't seem to make sense or that you know are completely wrong. This improves your odds of guessing the correct answer. If, for example, you can eliminate three choices, you have a 50/50 chance of guessing the correct answer. Otherwise, if you can't eliminate any choices, you have only a 20 percent, rather than a 50 percent, chance of getting the answer correct. The penalty for an incorrect answer is one-quarter point, so it may be advisable to guess.

The AP Physics test is a very difficult test, since you will more than likely encounter material that you've not studied before and because of the amount of work required to complete the test in a comparatively short period of time. However, with diligent studying, careful preparation, and a positive attitude, you can help yourself succeed.

Good Luck!

AP PHYSICS STUDY PLAN

When you begin to study for the AP Physics exam, the most important thing you should have is a plan. The first thing you should do is estimate how much time you have before the exam. The more time, the better. If, however, you're somewhat short on time, this study plan will be extremely valuable for you. We offer these different study plans to help maximize your time and studying. The first is a 12-Week Plan, which involves concentrated studying and a focus on the sample test results. The second is the more leisurely 24-Week Plan, the one that's favored by schools. Finally, if time is running short, you should use the Panic Plan. We don't want you to really panic—this plan is supposed to help you conquer that panic and help you organize your studying so that you can get the most out of your review work and still be as prepared as possible.

These plans are supposed to be flexible and are only suggestions. Feel free to modify them to suite your needs and your own study habits. But start immediately. The more you study and review the questions, the better your results will be.

THE 12-WEEK PLAN—2 LESSONS PER WEEK.

Week 1

Lesson 1— **Diagnostic Test**. The AP Physics Diagnostic Test is designed to help you determine what you need to know and where to focus your study. Take this test under simulated test conditions in a quiet room and keep track of the time it takes to complete the test. The test consists of three sections: General Physics, Mechanics, and Electricity and Magnetism. Each section consists of fourteen multiple choice questions and one free-response question. Regardless of which specific test you intend to take, you should answer all of the questions on this test to get an idea of your weakest areas.

Lesson 2— **Diagnostic Test—Answers**. Once you have completed the test, carefully check all of your answers, and read through the explanations. This may take quite a bit of time, as will all of the tests, but it will enable you to select those subject areas that you should focus on and spend the most amount of time studying. Armed with this information, you can start reviewing the chapters in the rest of the book. If you're short on study time, use the results of this test to focus your study efforts on the specific chapters in the review section that will help you understand the material that you missed on the test.

Week 2

Lesson 1— **Chapter One: Kinematics.** The review section of this book consists of eighteen chapters. It's an enormous amount of work, so you'll have to be extremely diligent about reviewing this material. These chapters fall under five major sections: Newtonian Mechanics, Thermal Physics, Electricity and Magnetism, Waves and Optics, and Atomic and Nuclear Physics.

Kinematics is the first chapter under the Newtonian Mechanics section, and 50 percent of the C-Level test consists of Newtonian Mechanics questions. Take your time to read through the first chapter. Underline or use a marker to highlight those areas that are unclear to you.

Lesson 2— **Chapter Two: Newton's Law of Motion**. Again, read through this chapter, mark whatever is unclear, and go back and read the material again, if necessary.

Week 3

Lesson 1— **Chapter Three: Work, Energy, Power**. As you continue your lessons, try to study in a quiet room, uninterrupted by others in your household or the TV, radio, or any outside noises.

Lesson 2— **Chapter Four: System of Particles, Linear Momentum**. Again, read through this chapter, mark whatever is unclear, and go back and read the material again, if necessary.

Week 4

Lesson 1— **Chapter Five: Circular Motion and Rotation**. You can, of course, break these lessons into sections. Work on half the chapter in the morning and the other half in the afternoon.

Lesson 2— **Chapter Six: Oscillations and Gravitations**. Read through this chapter, mark whatever is unclear, and then go back and read the material again, if necessary. You can always ask your teacher for additional information if you're having difficulty.

If you are taking the C-Level exam, you might want to spend the next week reviewing these six chapters.

WEEK 5

Lesson 1— **Chapter Seven: Temperature and Heat**. This is the first of two chapters under the Thermal Physics section.

Lesson 2— **Chapter Eight: Kinetic Theory and Thermodynamics**. If you find that you've finished reading and reviewing Temperature and Heat with time to spare, you can double up and complete Chapter 8. You will then have more time to reread these two chapters and go to the next lesson.

WEEK 6

Lesson 1— **Chapter Nine: Electrostatics**. The third unit is Electricity and Magnetism, and this material in the next five chapters represents 50 percent of the C-Level exam, so it pays to focus heavily on these chapters.

Lesson 2— **Chapter Ten: Conductors, Capacitors, Dielectrics.** This is the second chapter in this unit. Take your time to make sure you fully understand all of the material.

WEEK 7

Lesson 1— **Chapter Eleven: Electric Currents**. This is the midway point of this unit if you're preparing only for the C-Level exam. These questions on Electricity and Magnetism also represent at least 25 percent of the B-Level exam, so it's important to understand what you're studying.

Lesson 2— **Chapter Twelve: Magnostatics.** Again, if you find yourself finished with a section faster than you anticipated, or the pace of this study plan is too slow, feel free to add additional reading to your lessons.

WEEK 8

Lesson 1— **Chapter Thirteen: Electromagnetism**. This is the last chapter of this unit, and if you're taking the C-Level exam, you have several choices. You can either reread the material in the two major units that are covered on your exam, you can skip to the final tests given at the end of this book, or you can continue to read through the rest of the chapters to make sure you have a complete understanding of AP Physics.

Lesson 2— **Chapter Fourteen: Wave Motion.** This chapter is part of the Waves and Optics unit that consists of three chapters.

WEEK 9

Lesson 1— **Chapter Fifteen: Physical Optics**. If you find these chapters difficult, you might want to take a break from your reading. Give yourself a day or two to just relax – assuming you have the time.

Lesson 2— **Chapter Sixteen: Geometric Optics.** This has been a very concentrated period of study, and you're almost done. Make sure to keep highlighting anything you don't fully understand.

WEEK 10

Lesson 1— **Chapter Seventeen: Atomic Physics and Quantum Effects**. This chapter is the first in the Atomic and Nuclear Physics unit. These last two chapters represent about 15 percent of the B-Level test.

Lesson 2— **Chapter Eighteen: Nuclear Physics.** The end of a long study road. This is the last review chapter. If you have time to spare when you've completed all of these chapters, you might want to go back and check any topics or items that you didn't understand, and make an appointment with your teacher to go over these topics.

WEEK 11

Lesson 1— **AP Physics Practice Test 1, Level B**. Take this test and answer all of the questions you can, and then guess at those you don't know. Circle those questions that you guessed at so that you can zero in on those specific answers and so that you don't delude yourself into thinking that you really knew those answers in the first place. It's important to evaluate what you know. Check all of your answers.

Lesson 2— **AP Physics Practice Test 2, Level B**. Take this test and answer all of the questions you can, and then guess at those you don't know. Circle those questions that you guessed at so that you can zero in on those specific answers. Check all of your answers.

WEEK 12

Lesson 1— **AP Physics Practice Test 1, Level C**. Take this test and answer all of the questions you can, and then guess at those you don't know. Circle those questions that you guessed at so that you can zero in on those specific answers and so that you don't delude yourself into thinking that you really knew those answers in the first place. It's important to evaluate what you know. Check all of your answers.

Lesson 2— **AP Physics Practice Test 2, Level C**. Take this test and answer all of the questions you can and then guess at those you don't know. Circle those questions that you guessed at so that you can zero in on those specific answers. Check all of your answers.

THE 24-WEEK PLAN—1 LESSON PER WEEK

If you're lucky enough to have the extra time, the 24-Week Plan will enable you to better utilize your study time. You will now be able to spread out your plan into one lesson a week. This plan is ideal because you are not under any pressure, and you can take more time to review the material in each of the chapters. You will also have enough time to double-check the answers to those questions that might have given you problems. Keep in mind that the basis for all test success is practice, practice, practice. If you're taking the C-Level exam, you can either shorten your study time or take the extra few weeks to reread everything.

THE PANIC PLAN

We hate using this term, and while we hope you don't fall into this category, not everyone has the luxury of extra time to prepare for the AP Physics test. Perhaps, however, we can offer you a few helpful hints to get you through this period.

Read through the official AP Physics bulletin and this *AP Physics B/C Success* book and memorize the directions. One way of saving time on this, or any, test, is to be familiar with the directions in order to maximize the time you have to work on the questions. On this test, the directions are pretty simple. You may also want to take the time to look at additional material on the Internet. You can find more information about the AP Physics test on the Internet at www.collegeboard.org/ap/physics/.

Read the introduction to this book. It will be helpful in preparing for the test and give you an understanding of what you can expect on the exam and how much time you will have to complete both sections of the test.
Take the Diagnostic Test as well as the practice tests.
Focus whatever time you have left on those specific areas of the test that gave you the most difficulty when you took the practice tests. Whatever time you have before the exam, keep in mind that the more you practice, the better you will do on the final exam.

Physics Formulas
TABLE OF INFORMATION

Constants and Conversion Factors

1 unified atomic mass unit.	$1\ u = 1.66 \times 10^{-27}\ kg = 931\ MeV / c^2$
Proton mass	$m_p = 1.67 \times 10^{-27}\ kg$
Neutron mass	$m_n = 1.68 \times 10^{-27}\ kg$
Electron mass	$m_e = 9.11 \times 10^{-31}\ kg$
Magnitude of electron charge	$e = 1.60 \times 10^{-19}\ c$
Avogadro's number	$N_0 = 6.02 \times 10^{23}\ mol^{-1}$
Universal gas constant	$R = 8.31\ J / (mol \times K)$
Boltzmann's constant	$k_B = 1.38 \times 10^{-23}\ J/K$
Speed of light	$c = 3.00 \times 10^8\ m/s$
Planck's constant	$h = 6.63 \times 10^{-34} J \times s = 4.14 \times 10^{-15}\ eV \times s$
	$hc = 1.99 \times 10^{-25} J \times m = 1.24 \times 10^3\ eV \times nm$
Vacuum permittivity	$e_0 = 8.85 \times 10^{-12} C^2/N \times m^2$
Coulomb's law constant	$k = 1/4\pi e_0 = 9.0 \times 10^9 N \times m^2/C^2$
Vacuum permeability	$m_0 = 4p \times 10^{-7} Wb/(A \times m)$
Magnetic constant	$k' = m_0/4\pi \times 10^{-7} Wb/(A \times m)$
Universal gravitational constant	$G = 6.67 \times 10^{-11}\ m^3/kg \times s^2$
Acceleration due to gravity at the Earth's surface	$g = 9.8\ m/s^2$
1 atmosphere pressure	$1\ atm = 1.0 \times 10^5\ N/m^2 = 1.0 \times 10^5\ Pa$
1 electron volt	$1\ eV = 1.60 \times 10^{-19} J$
1 angstrom	$1\ Å = 1 \times 10^{-10}\ m$

Units

Name	Symbol
meter	m
kilogram	kg
second	s
ampere	A
kelvin	K
mole	mol
hertz	Hz
newton	N
pascal	Pa
joule	J
watt	W
coulomb	C
volt	V
ohm	W
henry	H
farad	F
weber	Wb
tesla	T
degree Celsius	°C
electron-volt	eV

Prefixes

Factor	Prefix	Symbol
10^9	giga	G
10^6	mega	M
10^3	kilo	k
10^{-2}	centi	c
10^{-3}	milli	m
10^{-6}	micro	m
10^{-9}	nano	n
10^{-12}	pico	p

Values of Trigonometric Functions For Common Angles

Angle	Sin	Cos	Tan
0°	0	1	0
30°	1 / 2	$\sqrt{3}/2$	$\sqrt{3}/3$
37°	3 / 5	4 / 5	3 / 4
45°	$\sqrt{2}/2$	$\sqrt{2}/2$	1
53°	4 / 5	3 / 5	4 / 3
60°	$\sqrt{3}/2$	1 / 2	$\sqrt{3}$
90°	1	0	∞

Newtonian Mechanics

a = acceleration	F = force
f = frequency	h = height
J = impulse	K = kinetic energy
k = spring constant	l = length
m = mass	N = normal force
P = power	p = momentum
r = radius or distance	s = displacement
T = period	t = time
U = potential energy	v = velocity or speed
W = work	x = position

ANSWER SHEET FOR DIAGNOSTIC TEST

General Physics

1 ⊂A⊃ ⊂B⊃ ⊂C⊃ ⊂D⊃ ⊂E⊃
2 ⊂A⊃ ⊂B⊃ ⊂C⊃ ⊂D⊃ ⊂E⊃
3 ⊂A⊃ ⊂B⊃ ⊂C⊃ ⊂D⊃ ⊂E⊃
4 ⊂A⊃ ⊂B⊃ ⊂C⊃ ⊂D⊃ ⊂E⊃
5 ⊂A⊃ ⊂B⊃ ⊂C⊃ ⊂D⊃ ⊂E⊃
6 ⊂A⊃ ⊂B⊃ ⊂C⊃ ⊂D⊃ ⊂E⊃
7 ⊂A⊃ ⊂B⊃ ⊂C⊃ ⊂D⊃ ⊂E⊃
8 ⊂A⊃ ⊂B⊃ ⊂C⊃ ⊂D⊃ ⊂E⊃
9 ⊂A⊃ ⊂B⊃ ⊂C⊃ ⊂D⊃ ⊂E⊃
10 ⊂A⊃ ⊂B⊃ ⊂C⊃ ⊂D⊃ ⊂E⊃
11 ⊂A⊃ ⊂B⊃ ⊂C⊃ ⊂D⊃ ⊂E⊃
12 ⊂A⊃ ⊂B⊃ ⊂C⊃ ⊂D⊃ ⊂E⊃
13 ⊂A⊃ ⊂B⊃ ⊂C⊃ ⊂D⊃ ⊂E⊃
14 ⊂A⊃ ⊂B⊃ ⊂C⊃ ⊂D⊃ ⊂E⊃
15 ⊂A⊃ ⊂B⊃ ⊂C⊃ ⊂D⊃ ⊂E⊃

Mechanics

1 ⊂A⊃ ⊂B⊃ ⊂C⊃ ⊂D⊃ ⊂E⊃
2 ⊂A⊃ ⊂B⊃ ⊂C⊃ ⊂D⊃ ⊂E⊃
3 ⊂A⊃ ⊂B⊃ ⊂C⊃ ⊂D⊃ ⊂E⊃
4 ⊂A⊃ ⊂B⊃ ⊂C⊃ ⊂D⊃ ⊂E⊃
5 ⊂A⊃ ⊂B⊃ ⊂C⊃ ⊂D⊃ ⊂E⊃
6 ⊂A⊃ ⊂B⊃ ⊂C⊃ ⊂D⊃ ⊂E⊃
7 ⊂A⊃ ⊂B⊃ ⊂C⊃ ⊂D⊃ ⊂E⊃
8 ⊂A⊃ ⊂B⊃ ⊂C⊃ ⊂D⊃ ⊂E⊃
9 ⊂A⊃ ⊂B⊃ ⊂C⊃ ⊂D⊃ ⊂E⊃
10 ⊂A⊃ ⊂B⊃ ⊂C⊃ ⊂D⊃ ⊂E⊃
11 ⊂A⊃ ⊂B⊃ ⊂C⊃ ⊂D⊃ ⊂E⊃
12 ⊂A⊃ ⊂B⊃ ⊂C⊃ ⊂D⊃ ⊂E⊃
13 ⊂A⊃ ⊂B⊃ ⊂C⊃ ⊂D⊃ ⊂E⊃
14 ⊂A⊃ ⊂B⊃ ⊂C⊃ ⊂D⊃ ⊂E⊃
15 ⊂A⊃ ⊂B⊃ ⊂C⊃ ⊂D⊃ ⊂E⊃

Electricity and Magnetism

1 ⊂A⊃ ⊂B⊃ ⊂C⊃ ⊂D⊃ ⊂E⊃
2 ⊂A⊃ ⊂B⊃ ⊂C⊃ ⊂D⊃ ⊂E⊃
3 ⊂A⊃ ⊂B⊃ ⊂C⊃ ⊂D⊃ ⊂E⊃
4 ⊂A⊃ ⊂B⊃ ⊂C⊃ ⊂D⊃ ⊂E⊃
5 ⊂A⊃ ⊂B⊃ ⊂C⊃ ⊂D⊃ ⊂E⊃
6 ⊂A⊃ ⊂B⊃ ⊂C⊃ ⊂D⊃ ⊂E⊃
7 ⊂A⊃ ⊂B⊃ ⊂C⊃ ⊂D⊃ ⊂E⊃
8 ⊂A⊃ ⊂B⊃ ⊂C⊃ ⊂D⊃ ⊂E⊃
9 ⊂A⊃ ⊂B⊃ ⊂C⊃ ⊂D⊃ ⊂E⊃
10 ⊂A⊃ ⊂B⊃ ⊂C⊃ ⊂D⊃ ⊂E⊃
11 ⊂A⊃ ⊂B⊃ ⊂C⊃ ⊂D⊃ ⊂E⊃
12 ⊂A⊃ ⊂B⊃ ⊂C⊃ ⊂D⊃ ⊂E⊃
13 ⊂A⊃ ⊂B⊃ ⊂C⊃ ⊂D⊃ ⊂E⊃
14 ⊂A⊃ ⊂B⊃ ⊂C⊃ ⊂D⊃ ⊂E⊃
15 ⊂A⊃ ⊂B⊃ ⊂C⊃ ⊂D⊃ ⊂E⊃

Diagnostic Test

SECTION I—GENERAL PHYSICS

Directions: Each question listed below has five possible choices. Select the best answer given the information in each problem, and mark the corresponding oval on the answer sheet.
(You may assume $g = 10$ m/s^2).

1. In Olde English measure, 4 fingers equal one palm, 2 spans equal one cubit, 3 feet equal one ell, 2 cubits equal one ell, and 3 palms equal one span. How many inches are there in one finger?

   (A) $\frac{4}{3}$

   (B) $\frac{3}{4}$

   (C) 16

   (D) $\frac{1}{16}$

   (E) 2

2. A calorimeter contains 200 g of ice at $-20^{\circ}C$. Heat is added to the system at the rate of 100 calories/s. In these units, the specific heats of ice, water, and steam may be taken as 0.5 cal/g–C°, 1.0 cal/g–C°, and 0.5 cal/g–C°, respectively. The heat of fusion of ice is 80 cal/g, and the heat of vaporization of water is 540 cal/g. Neglecting the specific heat of the calorimeter, describe quantitatively the state of the system at 920 s.

   (A) All steam
   (B) All water
   (C) All ice
   (D) 100g ice, 100g water
   (E) 100g water, 100g steam

3. Two polarizing sheets have their transmission directions arranged so that no light gets through. A third sheet is inserted between the two so that its transmission direction makes a 30° angle with the transmission direction of the first sheet. Unpolarized light of intensity I_o is incident on the first sheet. Find the intensity transmitted through the last sheet. Note that

   $$\sin 30° = \cos 60° = \frac{1}{2} \textit{ and } \sin 60° = \cos 30° = \frac{\sqrt{3}}{2}$$

   (A) None

   (B) $\frac{I_o}{8}$

   (C) $\frac{3I_o}{32}$

   (D) $\frac{I_o}{16}$

   (E) $\frac{\text{Log}_{10} I_o}{4}$

4. A string of length L and linear mass density μ is held taut by a force F exerted at either end. Find the time required for a transverse pulse to travel from one end of the string to the other.

(A) $2LF\mu$

(B) $\frac{F}{2\pi\mu L}$

(C) $\sqrt{F/\mu L}$

(D) $\frac{F}{\mu L}$

(E) $L\sqrt{\mu/F}$

5. A box of mass M starts from a state of rest on a table. The coefficient of kinetic friction between the box and the table is μ (where $\mu<1$). A cord attached to the side of the box passes over a pulley at the edge of the table and is connected to an equal mass M that hangs a distance, L, above the floor. If static friction is sufficiently small that the system starts to move, how long will it take the hanging mass to hit the ground?

(A) $2\sqrt{\frac{L}{g(1-\mu)}}$

(B) $\sqrt{\frac{L}{g\mu}}$

(C) $\frac{4L^2M}{g^2\mu^2}$

(D) $\sqrt{\frac{2L}{g}}$

(E) $\frac{M^2L^2}{2\mu g}$

6. A pith ball of mass m has a positive charge of q on its surface. The ball is thrown vertically upward with an initial speed of v_o in a uniform vertically downward electric field with magnitude E. How high does the ball go? Neglect air resistance, but do not neglect gravity.

(A) $\frac{v_o^2}{2g}$

(B) $\frac{qE}{v^o}$

(C) $\frac{v_o}{qmE}$

(D) $\frac{mv_o^2}{2(qE+mg)}$

(E) $\sqrt{\frac{3mEqv_o}{g}}$

7. A $+2\mu C$ charge is at point (6m, 0), and a $-8\mu C$ charge is at point (2m, 0). Find the coordinates of a point (not at infinity) where the electric field is zero.

(A) (10m, 0)

(B) $\left(\frac{14}{3}m,0\right)$

(C) $\left(\frac{26}{5}m,0\right)$

(D) (–3m, 0)

(E) (4m, 2m)

8. A string is passed over a frictionless pulley suspended from the ceiling, with a 3kg mass suspended from one end of the string and a 2kg mass at the other end. The 2kg mass starts out at floor level, and the 3kg mass starts some distance above the floor. If the system is released from rest, the 3kg mass hits the floor with a speed of 6 m/s. Find the initial distance of the 3kg mass from the floor. For simplicity, assume that g=10 m/s^2.

   (A) 2m
   (B) 5m
   (C) 9m
   (D) 20m
   (E) 32.5m

9. Suppose that in a given location on the earth's surface the earth's magnetic field has a downward component of 5x10^{-5}*T*, a northward component of 1x10^{-5}*T*, and no east-west component. What are the magnitude and direction of the force on a 4m length of wire that carries 200*A* horizontally, from S to N?

   (A) 0.008*N*, North
   (B) 0.04*N*, West
   (C) $.008\sqrt{26}N$, East
   (D) 0.008*N*, South
   (E) 0.04*N*, East

10. A merry-go-round of radius *R* rotates at constant speed with period *T*. What is the minimum coefficient of static friction between the merry-go-round and a box of mass *m* placed at its edge that will enable the box to remain on the surface without sliding?

   (A) $\frac{2\pi R}{T}$
   (B) $\frac{4\pi^2 R}{gT^2}$
   (C) $\frac{gT^2}{2R}$
   (D) Zero
   (E) $\frac{mgTR}{3}$

11. Two trains are following one another at slightly different speeds—the one in front at speed v_1 and the one trailing behind at speed v_2, where $v_1 > v_2$. Both sound their horns at the same frequency, *f*. Take the speed of sound in air as *v*, and calculate the beat frequency of the combined sounds for a passenger in the trailing train.

   (A) $f\left(\frac{v - v_1}{v - v_2}\right)$
   (B) $f\left(\frac{v + v_2}{v + v_1}\right)$
   (C) $f\left(\frac{v_1 - v_2}{v_1 + v_2}\right)$
   (D) $f\left(\frac{v_1 - v_2}{v + v_1}\right)$
   (E) $f\frac{2v}{v_1 + v_2}$

12. The radius of a newly discovered planet is half the radius of the earth, and its mass is one tenth of the mass of the earth. If an object weighs 200*N* on the surface of the earth, what will it weigh on the surface of the other planet?

(A) 20*N*
(B) 5*N*
(C) 2000*N*
(D) 10*N*
(E) 80*N*

13. In the circuit shown below, the current through the 80Ω resistor is 2A, the voltage across the 180Ω resistor is 240V, and the battery has an EMF of 440V and an unknown internal resistance, *r*. Find the value of the resistance, *X*.

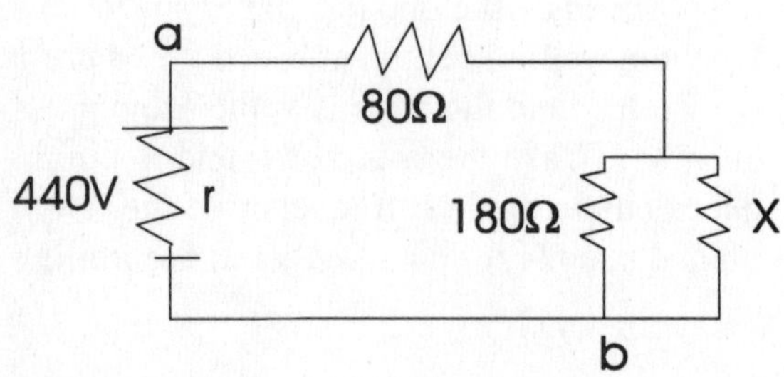

(A) 180Ω
(B) 220Ω
(C) 360Ω
(D) 440Ω
(E) 60Ω

14. A 0.25kg object is connected to a spring of spring constant $k = 25N/m$ and is set into oscillation with an initial spring potential energy of 12J and an initial kinetic energy of 4J. At what displacement are the kinetic and potential energies equal?

(A) 0.8m
(B) 1.0m
(C) 1.25m
(D) 2.0m
(E) 16m

SECTION II—FREE RESPONSE

Directions: Answer the following free-response question. Each question is designed to take approximately 15 minutes to answer. Note that each part within a question may not have equal weight. Show all work to obtain full credit, and avoid leaving important work on the green insert. Assume $g = 10\text{m/s}^2$.

15. A cord is used to vertically lower a block of mass, *M,* a distance, *d,* at a constant downward acceleration of $\frac{g}{3}$. In terms of *M, g,* and *d,*

    (a) find the tension in the cord.
    (b) find the work done by the cord on the block.
    (c) find the work done by gravity.
    (d) what is the change in kinetic energy of the block?

SECTION I—MECHANICS

1. A moving body of mass m makes a perfectly inelastic collision with a second body of twice its mass, initially at rest. Find the fraction of the initial kinetic energy that is lost in the collision.

(A) None
(B) One third
(C) One half
(D) Two thirds
(E) All

2. An object of mass 2kg makes an elastic collision with another object at rest and continues to move in the original direction but with one fourth of its original speed. What is the mass of the struck object?

(A) 0.5 kg
(B) 1.2 kg
(C) 2 kg
(D) 3.4 kg
(E) 8 kg

3. A boy throws a ball into the air as hard as he can and then bicycles as fast as he can, always staying under the ball in order to catch it. Assume the throw is from ground level. If the initial speed of the ball is 20m/s and the boy's cycling speed is 10m/s, find the time of flight. Take g as $10 m/s^2$, and note that $\sin 30° = \cos 60° = 0.5$ and $\sin 60° = \cos 30° = 0.866$.

(A) 3.46s
(B) 5.00s
(C) 6.28s
(D) 7.50s
(E) 8.87s

4. Assume that NASA wishes to send a manned spaceship to explore a large asteroid, at a distance of 1×10^{10}m from the earth. To approximately simulate earth gravity, NASA intends to accelerate the spaceship (from rest) at $10 m/s^2$ for the first half of the trip, then give the ship an equal negative acceleration for the remainder. However, the astronauts forget to reverse the engines until they cover 80% of the distance. Assuming that the astronauts arrive at the asteroid with zero velocity and that the reversed engines gave the ship a constant negative acceleration, what is the total elapsed time for the trip?

(A) 10,000 *s*
(B) 20,000 *s*
(C) 30,000 *s*
(D) 40,000 *s*
(E) 50,000 *s*

5. The three blocks in the figure below are released from rest and accelerate at the rate of $\frac{g}{10} m/s^2$, where g is the acceleration due to gravity. What is the coefficient of friction between the table and the horizontally moving block?

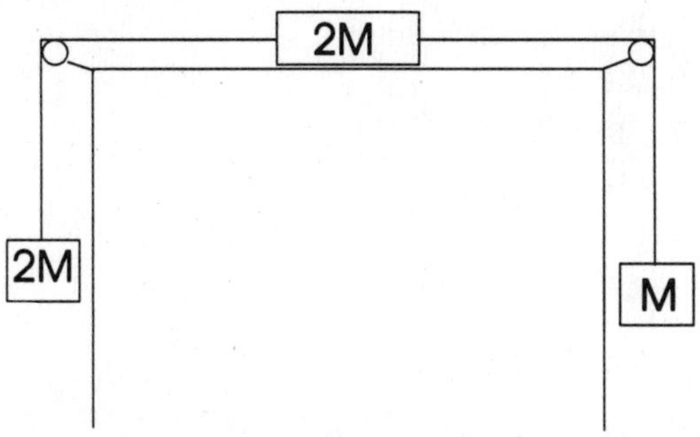

(A) 0.25
(B) 0.2
(C) 0.2 *MG*
(D) 0.5
(E) 2

6. Suppose that the earth was to somehow expand to become a sphere four times its present radius, but the total mass of the earth stayed constant. What is the new escape velocity of a rocket from the surface of the earth, in terms of the old escape velocity v_o?

(A) No change
(B) $\sqrt{2}v_o$
(C) $\frac{v_o}{2}$
(D) $\frac{v_o}{4}$
(E) $2v_o$

7. A flywheel with diameter *D* is pivoted on a horizontal axis. A rope is wrapped around the outside of the flywheel, and a steady force, *F*, is exerted on the rope. It is found that (starting from rest) *L* meters of rope are unwound in *t* seconds. What is the moment of inertia of the flywheel?

(A) $\frac{LtF}{2\pi D}$
(A) $\frac{4FL^2t^2}{D}$
(C) $\frac{\pi D^2}{4LtF}$
(D) $\frac{3L^3F}{Dt^2}$
(E) $\frac{FD^2t^2}{8L}$

8. A 2kg block traveling at 10 *m/s* on a horizontal, frictionless table strikes and becomes fastened to the end of a spring without a loss of energy. The other end of the spring is fixed. If the spring is compressed 100 cm before the block momentarily stops, what is the period of the resulting simple harmonic motion?

(A) 2s
(B) 2π s
(C) $\frac{\pi}{5 s}$
(D) 1 s
(E) $\frac{\pi}{2 s}$

9. A particle of mass m slides down a frictionless circular track of radius R, starting from rest from a position horizontally across from the center of the circle, as shown in the figure. Find the magnitude of the force exerted by the track on m at point B.

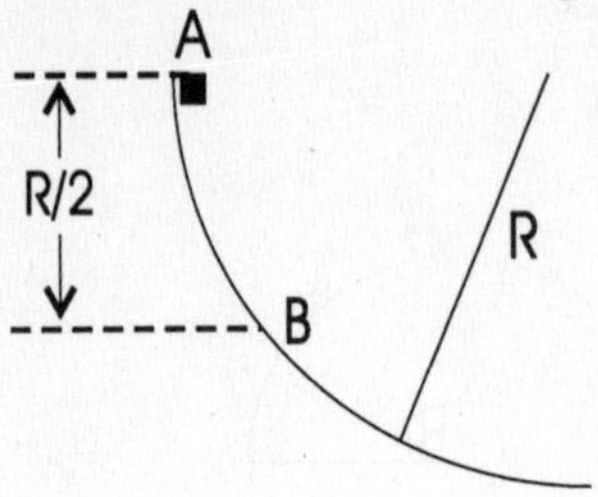

(A) 0

(B) $\frac{mg}{4}$

(C) $\frac{mg}{2}$

(D) mg

(E) $1.5mg$

10. An Atwood's machine consists of a pulley hanging from a ceiling, with unequal masses M_1 and M_2 ($M_1 > M_2$) attached to opposite ends of a rope hanging over the pulley. Assume that an Atwood's machine is set up on the surface of a planet. The pulley is a uniform solid disk ($I = MR^2/2$) of mass M kg and radius R, rotating on a frictionless axle. It is found that mass M_1 descends L meters in t seconds, starting from rest. No slippage occurs between the rope and the pulley. Find the acceleration due to gravity on the surface of the planet.

(A) $\frac{(M + 2M_1 + 2M_2)L}{(M_1 - M_2)t^2}$

(B) $\frac{(M + M_1 + M_2)R^2}{Lt^2}$

(C) $\frac{(2M + M_1 + M_2)L^2}{Rt^2}$

(D) $\frac{(M - M_1 - M_2)R}{4t^2}$

(E) $\frac{2L}{t^2}$

11. A stone is tied to a string of length R and whirled around in a vertical circle. Assuming that the energy remains constant, find the minimum speed it must have at the bottom of the circle in order for the string to remain taut at the top of the circle.

(A) $2\sqrt{gR}$

(B) $\sqrt{5gR}$

(C) $\sqrt{gR}$

(D) $2\pi\sqrt{gR}$

(E) $\frac{1}{\pi\sqrt{gR}}$

12. Corners *A, B*, and *C* of a right triangle are occupied by masses of 3, 8, and 6 *kg* respectively. Side *AC* is 3*m*, *BC* is 4*m*, and *AB* is 5*m*. If the magnitude of the net gravitational force exerted on the 6 *kg* mass by the other two masses is represented by *F*, and *G* is the universal gravitational constant, find the value of *F*.

(A) $\frac{5G\,kg^2}{m^2}$

(B) $\frac{6\sqrt{5}G\,kg^2}{m^2}$

(C) $\frac{\sqrt{13}G\,kg^2}{m^2}$

(D) $\frac{17G\,kg^2}{m^2}$

(E) 0

13. A small block of mass *m* slides down the frictionless loop-the-loop shown below, the circle having a radius *R*. The speed of the block as it passes a height, *h*, is *v*. Find the magnitude of the force that the track exerts on the block as it passes point *A* at the top of the loop.

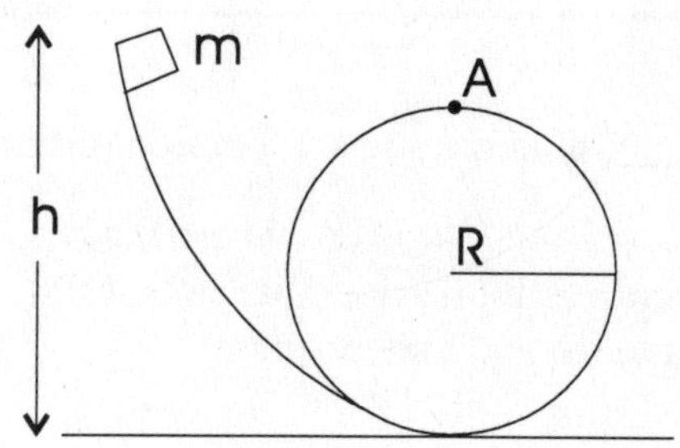

(A) $\frac{2mgh}{R-5mg}$

(B) $\frac{m\left(\frac{2v^2+gh^2}{R}\right)}{R}$

(C) $\frac{mv^2}{R}$

(D) $\frac{m(v^2+2gh-5gR)}{R}$

(E) mg

14. A dog running at a constant velocity of 11 *m/s* is 18 m behind its owner when the owner starts from a state of rest on a motor bike with a constant acceleration of 2 m/s^2. For a period of time, the dog will find itself ahead of its owner. How long is that time interval?

(A) 2s
(B) 3s
(C) 6s
(D) 7s
(E) 9s

SECTION II—FREE RESPONSE

Directions: Answer the following free-response question. Each question is designed to take approximately 15 minutes to answer. Note that each part within a question may not have equal weight. Show all work to obtain full credit, and avoid leaving important work on the green insert. Assume $g = 10\text{m/s}^2$.

15. The position of an object as a function of time is given by $x = 2t^3 - 24t - 18$, where t is in seconds and x is in meters. Displacements measured to the right are positive.

    (a) What is the average velocity of the object between $t = 1s$ and $t = 3s$?
    (b) What is its velocity at $t = 3$s?
    (c) At what time or times does the object stop?
    (d) What is its acceleration at each of the times in part (c)?

SECTION I—ELECTRICITY & MAGNETISM

1. Five equal positive charges, Q, are equally spaced on a semicircle of radius R as shown in the following diagram. What is the magnitude of the Coulomb force on a positive charge q located at the center of the semicircle?

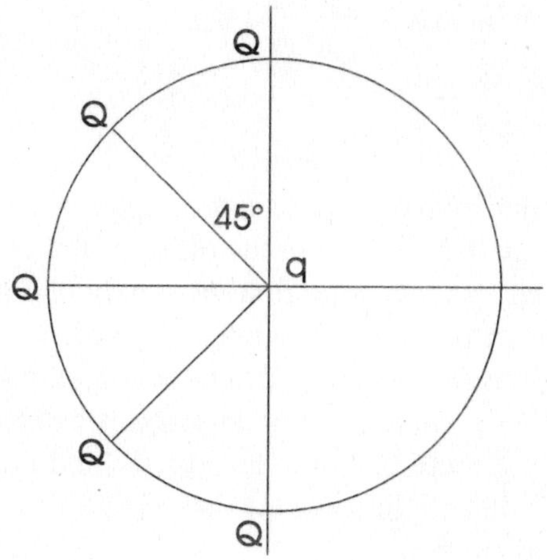

(A) Zero

(B) $\dfrac{kQq}{R^2}$

(C) $\dfrac{5kQq}{R^2}$

(D) $\dfrac{\sqrt{3}kQq}{R^2}$

(E) $\dfrac{\left(1+\sqrt{2}\right)kQq}{R^2}$

2. A positive point charge of magnitude Q is placed at the origin, and an unknown charge is placed at point $(a, 0)$. It is found that the electric field is zero at $(2a,0)$. Find the field at $(3a, 0)$.

(A) $\dfrac{kQ}{12a^2}$

(B) $\dfrac{7kQ}{144a^2}$

(C) Zero

(D) $\dfrac{3kQ}{22a^2}$

(E) $\dfrac{2kQ}{a^2}$

3. A thin rod stretches along the x-axis from $(2m, 0)$ to $(6m, 0)$ and has a uniform charge per unit length of 3 $\mu C/m$ distributed along its length. Find the potential at the origin. Use k=9x10^9 N-m^2/C^2.

(A) 6.75x10^3 V
(B) 2.7x10^4 ln(3) V
(C) 9x10^3 V
(D) Zero
(E) 0.025 V

4. The charge density on the surface of a conducting sphere of radius R is σ. A positive charge, q, of mass m is released from rest at a point outside the sphere at distance, "a," from the center of the sphere. Find its speed at the instant it is at distance, "b," from the center.

(A) $\sqrt{\dfrac{8k\sigma\pi R^2 q(b-a)}{mab}}$

(B) $\dfrac{2kq\sigma Ra}{mb}$

(C) Zero

(D) $qmk\ \ln\left(\dfrac{\sigma a}{(R^3 b)}\right)$

(E) $\dfrac{4\sigma ke^2}{abR}$

5. When a voltage, V, is placed across a resistor of resistance R, the power generated is 20W. The resistor is then snipped into three equal pieces—two of the pieces are combined in parallel and the third in series with that combination. If the same voltage, V, is placed across this combination, what will be the total power generated?

(A) 10W
(B) 26.7W
(C) 30W
(D) 40W
(E) 60W

6. The current in a 5Ω resistor varies with time according to the relation $i = 3t^2 - 4$, where i is in amperes and t is in seconds. Consider a time interval from $t = 1$s to t = 5s. What constant current would transport the same charge in that time interval as the time-varying current?

(A) 4A
(B) 9A
(C) 8 ln(2) A
(D) 15A
(E) 26A

7. A circular loop of wire of radius 2m and resistance 8Ω, located in the plane of the paper, is placed in a magnetic field perpendicular to the area of the wire. The magnetic field through the loop varies with time according to $B = 6t - t^2$, where t = time in seconds and the positive direction is into the paper. Find the magnitude and direction of the current flow in the loop at t=1s.

(A) Zero, no direction
(B) 2π A, clockwise
(C) 2π A, counterclockwise
(D) 4A, clockwise
(E) 4A, counterclockwise

8. A set of axes is laid out with the positive y-axis pointing toward the north and the positive x-axis pointing east. A long, straight wire carries a current of 12 amp along the y-axis toward the north. Find the magnitude and direction of the force on a +3μC charge moving due north with a velocity of 500 m/s, at the instant it passes through the point (6m, 2m). The magnetic constant is $k = \dfrac{\mu_o}{4\pi} = 10^{-7}\ N/A^2$.

(A) $6\text{x}10^{-10}N$, west
(B) $4\text{x}10^{-10}N$, north
(C) $6\text{x}10^{-10}N$, vertically up
(D) $4\text{x}10^{-10}N$, west
(E) $1\text{x}10^{-9}N$, vertically down

9. A positive point charge $+75\mu C$ is located on the y-axis at the point ($0m$, $-4m$), and a negative point charge - $50\mu C$ is located on the y-axis at ($0m$, $4m$). Find the y coordinates of all locations on the y-axis (not including ∞) at which the potential is zero.

   (A) $\frac{4(\sqrt{3}-\sqrt{2})}{(\sqrt{3}+\sqrt{2})}m$ only
   (B) $0.8m$ only
   (C) $20m$ only
   (D) $0.8m$ and $20m$ only
   (E) All points between the two charges

10. A $20{,}000\Omega$ resistor is connected in series with a capacitor, and a $40V$ potential is suddenly applied to the combination. The charge on the capacitor rises to 25% of its final value in $2\mu s$. Find the capacitance of the capacitor.

   (A) $2\text{x}10^{-8}F$
   (B) $16\mu F$
   (C) $\frac{1\text{x}10^{-10}}{\ln\left(\frac{4}{3}\right)}F$
   (D) $6e^{-3}F$
   (E) $4\ pF$

11. A hollow rubber ball of inner radius a and outer radius b has a uniform charge density ρ distributed through the rubber. Find the electric field at a distance r from the center, where $a<r<b$.

   (A) $\frac{\rho}{4\pi\varepsilon_0 r^2}$
   (B) $\frac{\rho(b^2-a^2)}{\varepsilon_0 r^2}$
   (C) $\frac{\rho r}{(b-a)}$
   (D) $\frac{\rho \ln}{\left(\frac{b}{a}\right)}$
   (E) $\frac{\rho(r^3-a^3)}{3\varepsilon_0 r^2}$

12. For the circuit given below, calculate the power dissipated in the 4Ω resistor.

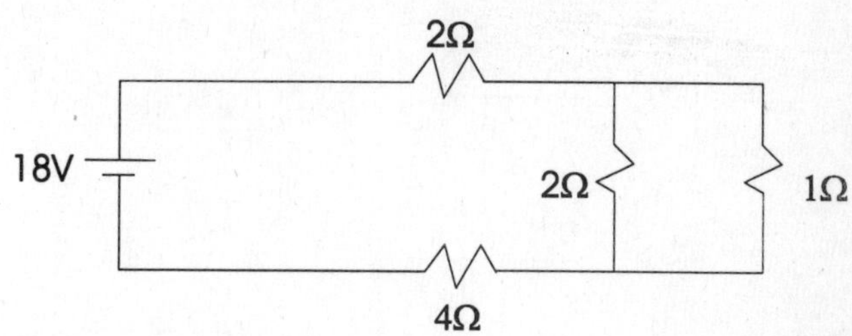

   (A) 81Ω
   (B) 36Ω
   (C) 29.16Ω
   (D) 64Ω
   (E) 3.25Ω

13. A charge, q, of mass m moves in a circular orbit of radius, R, perpendicular to a magnetic field of magnitude B. Find its kinetic energy.

(A) $\frac{q^2B^2R^2}{2m}$

(B) $\frac{mv^2}{R}$

(C) $q\vec{v}x\vec{B}$

(D) $\frac{mv}{qB}$

(E) $\frac{mB}{4\pi qR^2}$

14. A long, hollow, cylindrical shell of radius a, carrying a uniform negative surface charge density $-\sigma$, is surrounded by a coaxial cylindrical shell of radius b, carrying a surface charge of the same density but opposite sign. Find the electric field in terms of σ at a distance r from the axis of the cylinder, where $r>b$.

(A) Zero

(B) $\frac{\sigma(b-a)}{\varepsilon_o r}$

(C) $\frac{\sigma(b^2-a^2)}{\varepsilon_0 r^2}$

(D) $\frac{\sigma}{4\pi\varepsilon_o r^2}$

(E) $\frac{\sigma}{\varepsilon_0}$

SECTION II—FREE RESPONSE

Directions: Answer the following free-response question. Each question is designed to take approximately 15 minutes to answer. Note that each part within a question may not have equal weight. Show all work to obtain full credit, and avoid leaving important work on the green insert. Assume $g = 10\text{m/s}^2$.

15. In each of the following cases, determine the potential difference between points *x* and *y*, and state which point is at the higher potential. [Parts (a), (b), and (d) each show only a portion of the complete circuit.]

(a)

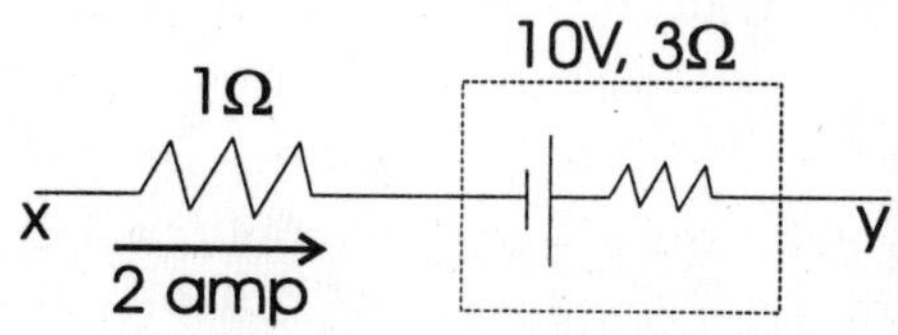

(b)

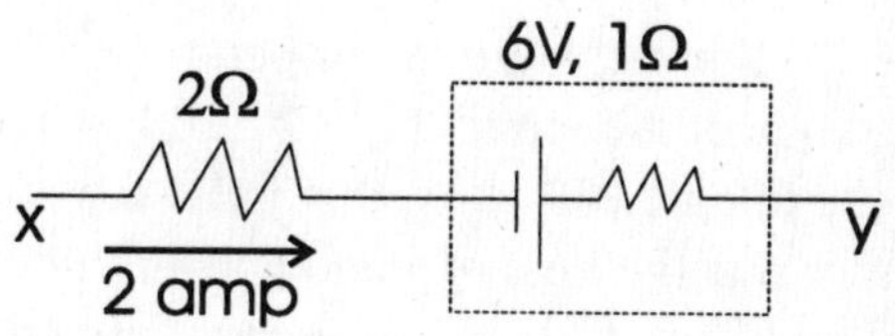

(c)

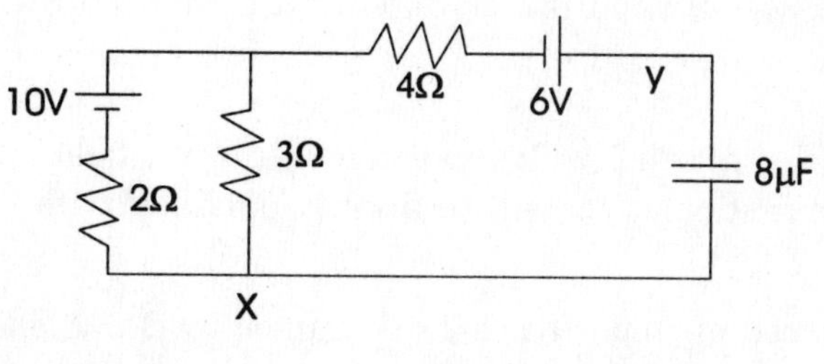

(d)

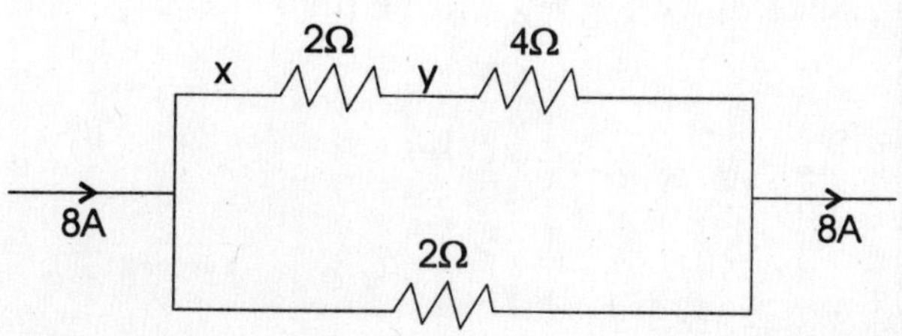

Diagnostic Test

ANSWERS AND EXPLANATIONS

SECTION I—GENERAL PHYSICS

QUICK-SCORE ANSWERS						
1. B	**3. C**	**5. A**	**7. A**	**9. B**	**11. D**	**13. C**
2. E	**4. E**	**6. D**	**8. C**	**10. B**	**12. E**	**14. A**

1. **The correct answer is (B).** Unit conversions should be done by cancellation:

(1finger)(1 palm/4 fingers)(1 span/3 palms)(1 cubit/2 spans)

(1 ell/2 cubits)(3 feet/1 ell) x (12inches/1 foot) = $\frac{3}{4}$ *inch*

2. **The correct answer is (E).** For a change in temperature, the heat supplied is given by $Q=mc\Delta T$. To heat the ice to 0°*C*, *Q*=(200*g*)(0.5 *cal/gC°*)(20*C°*)=2,000 *cal*. At the rate of 100 *cal/s*, this will take 20*s*. To melt the ice requires *Q=mL* where *L* is the heat of fusion. Then *Q*=(200*g*)(80 *cal/g*), requiring 16,000*cal* or 160*s*. To bring the water formed up to 100°*C* requires *Q*=(200*g*)(1 *cal/g-C°*)(100*C°*)=20,000*cal,* or another 200*s*. The elapsed time so far is 380*s*, leaving 920-380=540*s* for boiling. This will supply 54,000*cal*. At a heat of vaporization of 540 *cal/g*, this is sufficient to boil 100*g* of water.

3. **The correct answer is (C).** Starting with unpolarized light, the light transmitted through the first polarizer will be linearly polarized with intensity $\frac{I_o}{2}$. The succeeding intensities are determined by Brewster's Law: $I_{\text{final}} = I_{\text{initial}}\cos^2\theta$. The intensity through the second polarizer is then $\left(\frac{I_0}{2}\right)\left(\cos^2 30^0\right) = \frac{3I_0}{8}$, since $\cos 30^o = \sqrt{3}/2$. The final intensity is then

$$I = \left(\frac{3I_o}{8}\right)\cos^2 60° = \frac{3I_o}{32}, \text{ since } \cos 60° = \frac{1}{2}.$$

4. **The correct answer is (E).** The velocity of a transverse pulse traveling down a string is given by $v = \sqrt{F/\mu}$. The time required to travel a distance L is then $T = L\sqrt{\frac{\mu}{F}}$.

5. **The correct answer is (A).** Considering the hanging mass and taking downwards as positive, $Mg-T=Ma$, where T is the tension in the string. Considering the mass on the table, $f = \mu N = \mu Mg$. Then summing the horizontal forces, $T - \mu Mg = Ma$. Combining the first and last equations and solving for the acceleration, $\frac{g(1-\mu)}{2}$. Then, since $\frac{L = at^2}{2}$, the time is $t = 2\sqrt{\frac{L}{g(1-\mu)}}$.

6. **The correct answer is (D).** The forces on the pith ball are qE and mg, both downward. Taking upward as positive and using $\Sigma F = ma$, $-(qE + mg) = ma$, yielding $a = -\frac{(qe+mg)}{m}$. Using $v^2 = v_o^2 + 2ay$ (or by using energy conservation) and solving for y after substituting the value for acceleration, $y = \frac{mv_o^2}{2(qE+mg)}$

7. **The correct answer is (A).** Since electric fields are directed away from positive charges and toward negative charges, the point in question must be outside of the positive charges, closer to the smaller charge, and on the line joining them. Let that point have coordinates $(x, 0)$. Then,

$$\frac{k(2\mu C)}{(x-6m)^2} = \frac{k(8\mu C)}{(x-2m)^2}.$$

Canceling like factors, taking the square root of both sides, and solving for x, $x=10m$. The coordinates of the point are then $(10m, 0)$.

8. **The correct answer is (C).** The problem may be solved either by dynamics or by energy conservation. By the latter method, since the initial kinetic energy of the system is zero,
$(3kg)(10\ m/s^2)(x) = 0.5(3kg)(6m/s)^2 + 0.5(2kg)(6\ m/s)^2 + (2kg)(10\ m/s^2)(x)$, where x is the desired distance. Solving, $x = 9m$.

9. **The correct answer is (B).** The northward component of the earth's field will have no effect on a current directed northward. The force exerted by the downward component is given by $F = ILB \sin\theta = (200A)(4m)(5\text{x}10^{-5}T)\sin 90° = 0.04N$. The direction of the force, as determined by the right-hand rule, is toward the west.

10. **The correct answer is (B).** The coefficient of friction is $\mu = \frac{f}{N} = \frac{f}{mg}$.

Since $f = \frac{mv^2}{R}$, we have $v^2 = \mu gR$. But $v = \frac{2\pi R}{T}$. Substituting and

solving for μ, $\mu = \frac{4\pi^2 R}{gT^2}$.

11. **The correct answer is (D).** The wavelength behind the first train is the speed of sound relative to that train divided by the frequency sounded by that train, or

$\lambda = \frac{v + v_1}{f}$. The frequency heard by the second train is the velocity of sound relative to that train divided by the wavelength, or

$$f_2 = f\left(\frac{v + v_2}{v + v_1}\right).$$

Since $v_1 > v_2$, $f_2 < f$. The beat frequency is then the difference, $f - f_2$. Inserting the expression above for f_2 and simplifying, Beat frequency =

$$f\left(\frac{v_1 - v_2}{v + v_1}\right).$$

12. **The correct answer is (E).** The gravitational force exerted by a planet on a mass located on its surface is directly proportional to the planet's mass and inversely proportional to the square of its radius. Using subscripts *P* for the planet and *E* for the earth,

$$F_P = \left(\frac{M_P}{M_E}\right)\left(\frac{R_E}{R_P}\right)^2 F_E = \left(\frac{1}{10}\right)(2)^2(200N) = 80N \cdot$$

13. **The correct answer is (C).** The current through the 180Ω resistor is $\frac{240V}{180\Omega} = \left(\frac{4}{3}\right)A$. Since the current through the 80Ω resistor is 2*A*, this leaves $\left(\frac{2}{3}\right)A$ to go through the unknown resistor. The voltage across this resistor is 240*V*, so the resistance is given by $\frac{240V}{\left(\frac{2}{3}\right)A} = 360\Omega$

14. **The correct answer is (A).** The total energy is 16*J*, which is conserved. At the point where the kinetic and potential energies are equal, each will be 8*J*. The potential energy is given by $U = 0.5kx^2 = 0.5(25\ \frac{N}{m})x^2 = 8J$ from which $x = 0.8m$.

SECTION II—FREE RESPONSE

15. (a) Taking upwards as positive, and using $\Sigma F = Ma$,

$$T\text{-}Mg = \left(\frac{-g}{3}\right) \text{ or } T = \frac{2Mg}{3}.$$

(b) $W = Td\cos 180^\circ = \frac{-2Mgd}{3}$

(c) $W = Mgd\cos 0^\circ = Mgd$

(d) $DK = W_{total} = \frac{Mgd}{3}$

SECTION I—MECHANICS

QUICK-SCORE ANSWERS						
1. D	3. A	5. A	7. E	9. E	11. B	13. D
2. B	4. E	6. C	8. C	10. A	12. C	14. D

1. **The correct answer is (D).** From momentum conservation,

   mv=3mV, so $V=\frac{v}{3}$, where v = original velocity and V = final velocity.

   Then $K_f=\frac{1}{2}(3m)\left(\frac{v}{3}\right)^2=\frac{mv^2}{6}$ and K_{lost}=K_i–K_f = $\frac{mv^2}{3}$.

   The fraction lost is $\frac{K_{lost}}{K_i}=\frac{2}{3}$.

2. **The correct answer is (B).** From momentum conservation,

   $2v=2\left(\frac{v}{4}\right)+mV$, yielding $V=\frac{3v}{2m}$. From kinetic energy conservation,

   $\frac{1}{2}(2)v^2=\frac{1}{2}(2)\left(\frac{v}{4}\right)^2+\frac{1}{2}mV^2$. Substituting expression (1) for V and simplifying, m = 1.2 *kg*.

3. **The correct answer is (A).** The x-component of the ball's velocity must match the boy's velocity of 10 *m/s*. The initial velocity triangle is then a 30º–60º–90º triangle, yielding an initial vertical velocity of 17.3 *m/s*.

   Taking vertically upwards as positive and using $y=v_0t+\frac{1}{2}at^2$, with a = -10 m/s^2, and y=0, since the ball returns to ground level, "t" may be factored out, yielding t = 3.46 s.

4. **The correct answer is (E).** Using $\Delta x=v_0t+\frac{1}{2}at^2$ for the first leg, t = 4x10^5*s*. Using $v=v_o+at$, the velocity at the end of the first leg is 4x10^5 *m/s*. This is the initial velocity for the second leg. Using

   $\Delta x=\frac{(v_o+v_{final})t}{2}$ and $v_{final}=0$ for that leg, t = 1x10^4 *s*. The total time is then 5x10^4 *s*.

5. **The correct answer is (A).** For the block of mass *M*, $T_1 - Mg = \frac{Mg}{10}$, where T_1 is the tension in the right-hand rope and the positive direction is upward. This yields $T_1 = \frac{11}{10Mg}$. For the other hanging block, taking downward as positive, $2Mg - T_2 = \frac{Mg}{5}$. This yields $T_2 = \frac{9Mg}{5}$. For the block on the table, taking the positive direction to the left, $T_2 - T_1 - f = \frac{2Mg}{10}$. Substituting the tensions yields $f = \frac{Mg}{2}$. But $f=\mu N= \mu(2Mg)$. Thus $\mu=0.25$.

6. **The correct answer is (C).** Conserving energy, $K_I+U_I = K_f+U_f$.

$\frac{1}{2}mv_{esc}^2 - \frac{GMm}{R_E} = 0$, where M = mass of Earth and m = mass of rocket.

Then $v_{esc} = \sqrt{\frac{2GM}{R_E}}$, and, thus, if the radius of the Earth is quadrupled, the escape velocity will become half of its former value.

7. **The correct answer is (E).** Since $\Sigma\tau = I\alpha$, the moment arm is $R = \frac{D}{2}$, and $\alpha = \frac{a}{R}$, we have $\frac{FD}{2} = \frac{Ia}{\left(\frac{D}{2}\right)}$. Also, $L = \frac{at^2}{2}$. Eliminating *a* from the two equations, $I = FD^2t^2/8L$.

8. **The correct answer is (C).** From energy conservation, $\frac{1}{2}kx^2 = \frac{1}{2}mv^2$, from which $k = \frac{200N/m}{\text{m}}$. For a spring, $T = 2\pi\sqrt{\frac{m}{k}}$, yielding $T = \frac{\pi}{5}s$.

9. **The correct answer is (E).** Taking point *B* as the zero of potential energy and conserving energy between *A* and *B*, $\frac{mgR}{2} = \frac{mv^2}{2}$, so $v^2=gR$.

A radius drawn from B makes a 30° angle with the horizontal. The two forces on m are the normal force N and the weight mg. If the weight is broken into radial and tangential components, then in the radial direction,

$\Sigma F = N - mg\sin 30° = \frac{mv^2}{R}$. Using $v^2=gR$, we obtain $N = 1.5mg$.

10. **The correct answer is (A).** Applying $\Sigma F = ma$ to M_1, where T_1 is the tension in the rope connected to M_1, and taking downwards as positive, $M_1g - T_1 = M_1a$.
Similarly for M_2, taking upwards as positive, $T_2 - M_2g = M_2a$.

Using $\Sigma\tau = I\alpha$, where $\alpha = \frac{a}{R}$, $\quad T_1R - T_2R = \left(\frac{MR^2}{2}\right)\left(\frac{a}{R}\right)$.
Simplifying and adding the three equations yields $M_1g - M_2g = (0.5M + M_1 + M_2)a$
But $L = 0.5at^2$ from kinematics. Substituting and solving for g,

$$g = \frac{(M + 2M_1 + 2M_2)L}{(M_1 - M_2)t^2}.$$

11. **The correct answer is (B).** If v is the speed at the bottom and V is the speed at the top, then from energy conservation, $\frac{Mv^2}{2} = Mg(2R) + \frac{MV^2}{2}$.

The centripetal force at the top is supplied by gravity only: $Mg = \frac{MV^2}{R}$.

Eliminating V between the two equations, $v = \sqrt{5gR}$.

12. **The correct answer is (C).** The force F exerted by one mass on another is given by $F = \frac{GMm}{r^2}$. Then, the force exerted on the 6kg mass by the 3kg mass is $\frac{2Gkg^2}{m^2}$, and the force exerted on the 6kg mass by the 8kg mass is $\frac{3Gkg^2}{m^2}$. These forces are at right angles, so the net force is given by the Pythagorean theorem as $\frac{\sqrt{13}Gkg^2}{m^2}$.

13. **The correct answer is (D).** By conserving energy between height h and point A, and using V as the velocity at A,

$$\frac{1}{2}mv^2 + mgh = \frac{1}{2}mV^2 + mg(2R)\cdot$$

Using Newton's second Law at point A, $N + mg = \frac{mV^2}{R}$, where N is the normal force exerted by the loop. Eliminating V between the two equations, $N = \frac{m(v^2 + 2gh - 5gR)}{R}$.

14. **The correct answer is (D).** From kinematics, $x - x_o = v_o t + \frac{1}{2}at^2$.

For the dog, x=(11 m/s)t. For the owner, $x - 18m = \frac{(2m/s^2)t^2}{2}$. Eliminating x and solving the factorable quadratic equation that results, t = 2s or 9s. Consequently, there is a 7-second interval between the two times that the dog and the owner were at the same location.

SECTION II—FREE RESPONSE

15. Since $x = 2t^3-24t-18$, the position of the object is at $-36m$ at $t = 3s$ and at $-40m$ at t = $1s$. The average velocity is the displacement over the time interval, or $4m/2s = 2$ m/s.

(a) The instantaneous velocity is the derivative with respect to time of the position, yielding $v = 6t^2-24$. At $t = 3s$, $v = 30$ m/s.

(b) The velocity is zero when $6t^2-24=0$, yielding $t = \pm 2$ s. (Both answers are meaningful. The negative time simply means 2 s before the clock was started.)

(c) The acceleration is the time derivative of the instantaneous velocity, yielding $a = 12t$. At $t = 2s$, $a = 24$ m/s^2. At $t = -2s$, $a = -24$ m/s^2.

SECTION I—ELECTRICITY AND MAGNETISM

QUICK-SCORE ANSWERS						
1. E	3. B	5. D	7. C	9. D	11. E	13. A
2. B	4. A	6. E	8. A	10. C	12. C	14. B

1. **The correct answer is (E).** The forces exerted by the charges at the 90° and 270° positions are equal and opposite and cancel each other. The force exerted by the charge at 180° is $\frac{kQq}{R^2}$ toward the right. The forces exerted by the other two charges, each of magnitude $\frac{kQq}{R^2}$, are at right angles to each other and can be combined by the Pythagorean theorem into a single force $\frac{\sqrt{2}kQq}{R^2}$ to the right. (Alternatively, this can be accomplished by breaking the forces into components.) The resultant force is then of magnitude $\frac{\left(1+\sqrt{2}\right)kQq}{R^2}$.

2. **The correct answer is (B).** If the unknown charge is called q, the field at $(2a, 0)$ is given by $E = \frac{kQ}{(2a)^2} + \frac{kq}{a^2} = 0$, yielding $q = \frac{-Q}{4}$. The field at $(3a, 0)$ is then

$$E = \frac{kQ}{(3a)^2} + \frac{k\left(-\frac{Q}{4}\right)}{(2a)^2} = \frac{7kQ}{144a^2}.$$

3. **The correct answer is (B).** The potential dV due to a small piece of charge dq at distance x from the origin is $dV = \frac{kdq}{x} = \frac{k\lambda dx}{x}$ where $\lambda = 3\text{x}10^{-6} C/m$. Integrating from $x=2m$ to $x=6m$, and recalling that $\ln(6) - \ln(2) = \ln(6/2) = \ln(3)$, we have $V=2.7\text{x}10^4 \ln 3$ *Volts*.

4. **The correct answer is (A).** The amount of charge on the surface of the sphere is $Q=\sigma(4\pi R^2)$. At a point outside the sphere, a distance *r* from the center, the charge on the sphere creates a potential given by $V=\frac{kQ}{r}$. The potential energy of a point charge, *q*, is then $U=qV$. Conserving energy, $K_i + U_I = K_f + U_f$, where *i* and *f* stand for initial and final, respectively. Then,

$$0+\frac{k\sigma 4\pi R^2 q}{a}=\frac{1}{2}mv^2+\frac{k\sigma 4\pi R^2 q}{b}.$$ Solving for *v*,

$$v=\sqrt{\frac{8k\sigma\pi R^2 q(b-a)}{mab}}.$$

5. **The correct answer is (D).** Since $P=\frac{V^2}{R}$, the relationship between *V* and *R* is given by $20W=\frac{V^2}{R}$. The three resistors will each have resistance $\frac{R}{3}$. The parallel combination will then have resistance $\frac{R}{6}$, and the series combination will have a total resistance $\frac{R}{2}$. The power across that combination is then given by $P=\frac{V^2}{\frac{R}{2}}=\frac{2V^2}{R}=2(20W)=40W$.

6. **The correct answer is (E).** Current is given by $i=\frac{dq}{dt}$, so $q=\int_1^5(3t^2-4)dt=104C$. The constant current would then be $\frac{104C}{4s}=26A$.

7. **The correct answer is (C).** The flux at any instant is given by $\phi=\vec{B}\cdot\vec{A}=(6t-t^2)\pi(2)^2$ *Wb*, and the induced *EMF* is $\frac{-d\phi}{dt}=-(24\pi-8\pi t)$. At t=1s, the *EMF* is then -16π *Volts* and the

current is $I = \frac{-16\pi V}{8\Omega} = -2\pi A$. The current (as indicated by the negative sign or by application of Lenz's Law) will be counterclockwise, so as to create a magnetic field out of the paper to counteract the increasing flux into the paper.

8. **The correct answer is (A).** The field at a distance *r* from a long straight wire is $B = \frac{\mu_o I}{2\pi r}$. With the current going north, the field at (6*m*, 2*m*) will be vertically down. The force on a charge moving north will be directed west and will have magnitude $F = qvB = \frac{\mu_o Iqv}{2\pi r}$, where in this instance *r* = 6μ. Substituting,
   $F = 6x10^{-10}N$.

9. **The correct answer is (D).** At a point between the two charges with coordinates (0, *y*), the total potential is given by
   $$V = \frac{k(-50\mu C)}{4m - y} + \frac{k(75\mu C)}{y + 4m} = 0.$$
   Dividing through by 25k μ*C* and solving, *y*=0.8*m*. At a point with coordinates (0,*y*) located above the upper charge,
   $V = \frac{k(-50\mu C)}{y - 4m} + \frac{k(75\mu C)}{y + 4m} = 0$. Solving in a similar fashion, *y*=20*m*.
   There is no point below the lower charge at which the potential is zero.

10. **The correct answer is (C).** The charge on a capacitor being charged is given by

    $Q = Q_m(1 - e^{-\frac{t}{RC}})$, where Q_m is the maximum charge after a long period of charging. In this instance, $Q = \frac{Q_m}{4}$, so that $e^{-\frac{t}{RC}} = \frac{3}{4}$, yielding
    $$C = \frac{t}{R\ln\left(\frac{4}{3}\right)}.$$

    Substituting the given values, $C = \frac{1x10^{-10}}{\ln\left(\frac{4}{3}\right)}F$.

11. **The correct answer is (E).** Gauss' Law applies to a Gaussian sphere of radius *r* (where *a*<*r*<*b*) yields $\oint \vec{E} \cdot d\vec{A} = \frac{q}{\varepsilon_o}$.

$$E 4\pi r^2 = \frac{\rho\left(\frac{4}{3}\pi r^3 - \frac{4}{3}\pi a^3\right)}{\varepsilon_o}$$

Solving for *E*, $E = \frac{\rho(r^3 - a^3)}{3\varepsilon_o r^2}$.

12. **The correct answer is (C).** The parallel combination of 1Ω and 2Ω gives a combined resistance of $\left(\frac{2}{3}\right)\Omega$, which combines in series with the other resistors to give an equivalent resistance of $\left(\frac{20}{3}\right)\Omega$. Using *V*=*IR*, the current through the equivalent circuit is $\frac{18V}{\left(\frac{20}{3}\right)\Omega} = \left(\frac{27}{10}\right)A$, which is then the current through the 4Ω resistor. The power loss through that resistor is *P*=*I*²*R*, yielding $P = \left(\frac{2916}{100}\right)\Omega = 29.16\Omega$.

13. **The correct answer is (A).** The force on the particle is $f = qvB = \frac{mv^2}{R}$, from which $v = \frac{qBR}{m}$. Since $KE = \frac{mv^2}{2}$, we obtain $KE = \frac{q^2B^2R^2}{2m}$.

14. **The correct answer is (B).** Taking a Gaussian cylinder of radius *r* and length *L*, and applying Gauss' Law,

$$\oint \vec{E} \cdot d\vec{A} = \frac{q}{\varepsilon_o}$$

$E 2\pi r L = \frac{\sigma(2\pi bL - 2\pi aL)}{\varepsilon_o}$, from which $E = \frac{\sigma(b-a)}{\varepsilon_o r}$.

SECTION II—FREE RESPONSE

15. In diagram #1, V_{xy}=(2A)(3Ω)-10V+(2A)(1Ω) = -2V, indicating that point y is at the higher potential.

In diagram #2, V_{xy}=(2A)(1Ω)-6V+(2A)(2Ω) = 0V, indicating that point x is at the higher potential.

In diagram #3, no current goes through the capacitor branch. The 2Ω and 3Ω resistors are then in series, and the current through the left-hand loop is I=V/R = 10V/5Ω = 2A clockwise.

Then V_{xy} = -6V-(2A)(3Ω) = -12V, indicating that point y is at the higher potential.
In diagram #4, the total resistance of the top branch is 2Ω + 4Ω= 6Ω. The equivalent resistance of the parallel combination is then obtained from

$$\left(\frac{1}{R}\right)=\left(\frac{1}{6\Omega}\right)+\left(\frac{1}{2\Omega}\right),$$ so that R=1.5Ω .

The voltage across the combination is then V=(8A)(1.5Ω) = 12V, from which the current in the top branch is 12V/6Ω = 2A. The voltage V_{xy} is then V_{xy} =(2A)(2Ω) =4V, indicating that point x is at the higher potential.

UNIT 1
Newtonian Mechanics

Chapter 1
KINEMATICS

MOTION IN ONE DIMENSION

Motion in one dimension is exemplified by motion along a straight line.

DEFINITIONS

Coordinate System—Motion is a change of position, so in order to discuss motion, we must first discuss position. To discuss position, we must choose an origin and a reference direction. These choices are arbitrary and must be made before the position can be defined.

For example, the origin can be in the middle of a horizontal line, and the reference direction can be to the right. We will call the reference direction the "+ direction." In one dimension, only one other direction is possible. In this example, that direction is towards the left. We call that direction the "– direction."

When we have identified a reference direction and origin, we say that we have defined the (one dimensional) *coordinate system* that we will be using.

Position—The position of an object in one dimension is specified by stating two things: the distance from the origin and the direction from the origin to the object. Thus, an object with position –6 meters is 6 meters to the left of the origin. If it moves to a position of –3 meters, it is now 3 meters to the left of the origin.

Displacement—The change in position is calculated by subtracting the initial position from the final position. In the subtraction, the + and – direction signs are treated as algebra signs. An object which moves from –6 meters to –3 meters has a change of position of +3 meters.

The algebraic representation of this calculation uses x as the symbol for position and Δx as the symbol for change in position. We write:

$$\Delta x = x_{FINAL} - x_{INITIAL} = -3\,\text{m} - (-6\,\text{m}) = +3\,\text{m}$$

Δx is called the *displacement* of the object.

Velocity—Velocity is defined as the rate of change of position. Average velocity, $< v >$, is defined as the change in position, Δx, divided by the time required to make the change, Δt:

$$< v > = \frac{\Delta x}{\Delta t}$$

Instantaneous velocity is the limit of the average velocity as the time interval (and thus also the change in position) approaches zero. It is the time derivative of the position.

$$v = \lim_{\Delta t \to 0} \frac{\Delta x}{\Delta t} = \frac{dx}{dt}$$

Velocity can be positive or negative, depending on the direction of the displacement. The velocity can be zero while the position is not zero. The velocity can be large at a time when the position is zero.

Acceleration—Acceleration is defined as the rate of change of velocity. Average acceleration, $< a >$, is defined as the change in velocity, Δv, divided by the time required to make the change, Δt:

$$< a > = \frac{\Delta v}{\Delta t}$$

Instantaneous acceleration is the limit of the average acceleration as the time interval (and thus also the change in velocity) approaches zero. It is the time derivative of the position.

$$a = \lim_{\Delta t \to 0} \frac{\Delta v}{\Delta t} = \frac{dv}{dt}$$

Acceleration can be positive or negative, depending on the direction of the change in velocity. The acceleration can be zero while the velocity is not zero. The acceleration can be large at a time when the velocity is zero.

MOTION WITH CONSTANT ACCELERATION

EQUATIONS

Position as a Function of Time—For one dimensional motion with constant acceleration, a, the position, x, as a function of time, t, is given by

$$x = \frac{1}{2}at^2 + v_0 t + x_0$$

where x_0 is the position at time $t = 0$ (the initial position) and v_0 is the velocity at time $t = 0$ (the initial velocity). The position is the definite integral of the velocity from $t = 0$ to t.

Velocity as a Function of Time—For one dimensional motion with constant acceleration, a, the velocity, v, as a function of time, t, is given by

$v = at + v_0$, where v_0 is the velocity at time $t = 0$ (the initial velocity). The velocity is the definite integral of the (constant) acceleration from $t = 0$ to t. It is also the time derivative of the position given above.

Velocity as a Function of Position—The two equations give the complete solution for the motion of an object (for the case of constant acceleration). A partial solution can be created by combining those two so as to eliminate the time variable. The result gives the velocity as a function of position:

$$v^2 - v_0^2 = 2a(x - x_0)$$

This equation does not allow determination of the direction of the velocity since a velocity of either sign will satisfy it. This equation is replaced later, for any acceleration, by the work-energy theorem.

FREE "FALL"

It has been determined experimentally that any object falling without resistance near the surface of the earth has a downward acceleration of 9.8 m/sec^2. This acceleration is said to be due to the earth's gravity.

An object moving vertically and subject to only the earth's gravity is an ideal example of one dimensional motion with constant acceleration. It is important to note that the acceleration is the same whether the object is

a. moving upward (with decreasing magnitude of velocity),
b. moving downward (with increasing magnitude of velocity), or
c. standing still at the top of the path (zero velocity).

Problem solutions are typically set up with the origin at the lowest point in the problem and with $t = 0$ when the object begins its flight.

In many physics exercises, cars are understood to accelerate forward at a constant rate when the gas pedal is pressed. They are understood to accelerate backwards at a constant rate when the brake pedal is pressed.

GRAPHS

- If the position of a particle is plotted versus time, the slope of the position graph is the velocity of the particle.

The dashed line is tangent to the position curve at 1 seconds in the first graph on the next page. The slope of that line is about 2m/sec.

Calculus users, note that the slope of the graph is just the derivative of the position function with respect to time.

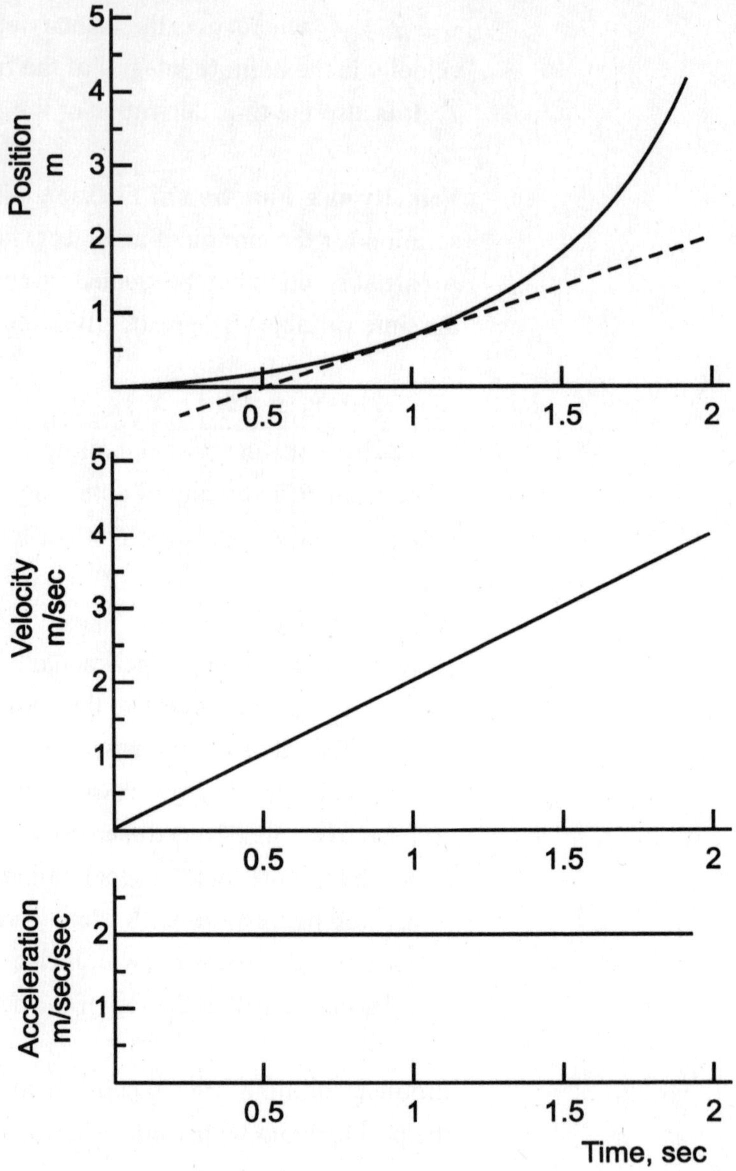

- If the velocity of a particle is plotted versus time, the slope of the velocity graph is the acceleration of the particle. In the graph, the slope of the velocity graph is constant, as is the acceleration.

Calculus users, note that the slope of the graph is just the derivative of the velocity function with respect to time.

- Calculus users, note that the definite integral of velocity from one time to another is the change in position of the particle represented by the area under the velocity curve on the graph. Also, the definite integral of acceleration from one time to another is the change in velocity of the particle, represented by the area under the acceleration curve on the graph.

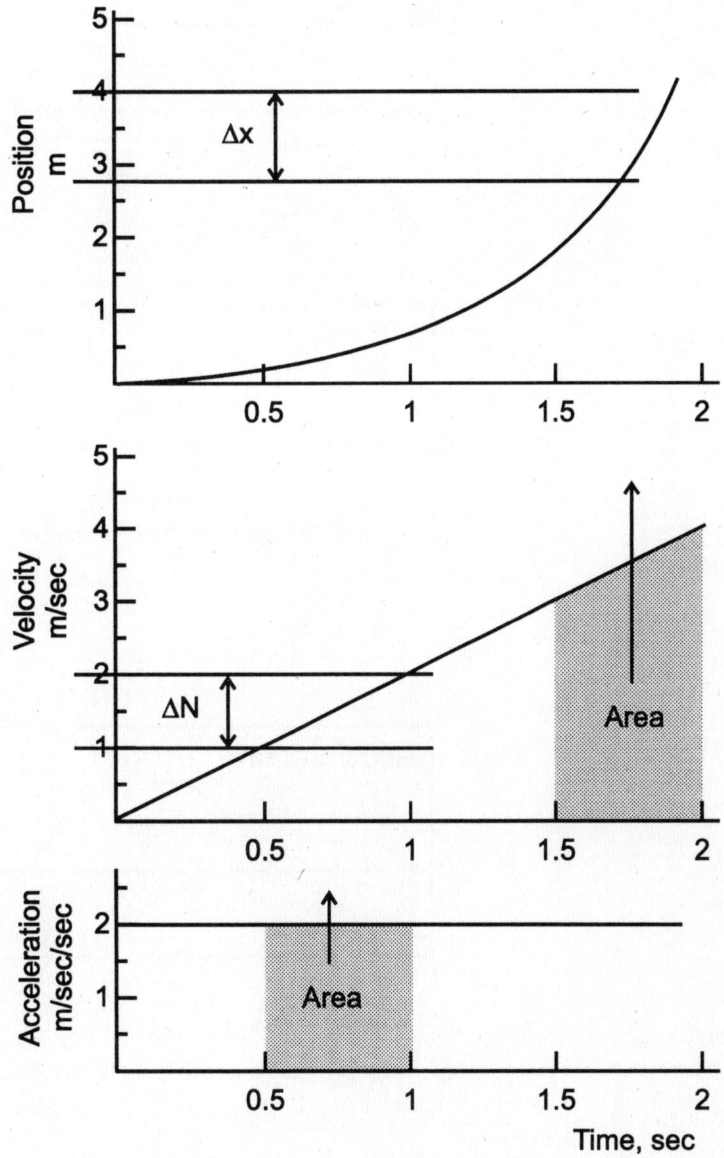

The area under the acceleration curve between .5 and 1.0 seconds is equal to the change in velocity during that time interval. The area under the velocity curve between 1.5 and 2.0 seconds is the change in position during that time. As an exercise, see how the slopes (derivatives) of position and velocity agree with velocity and acceleration in the graphs below.

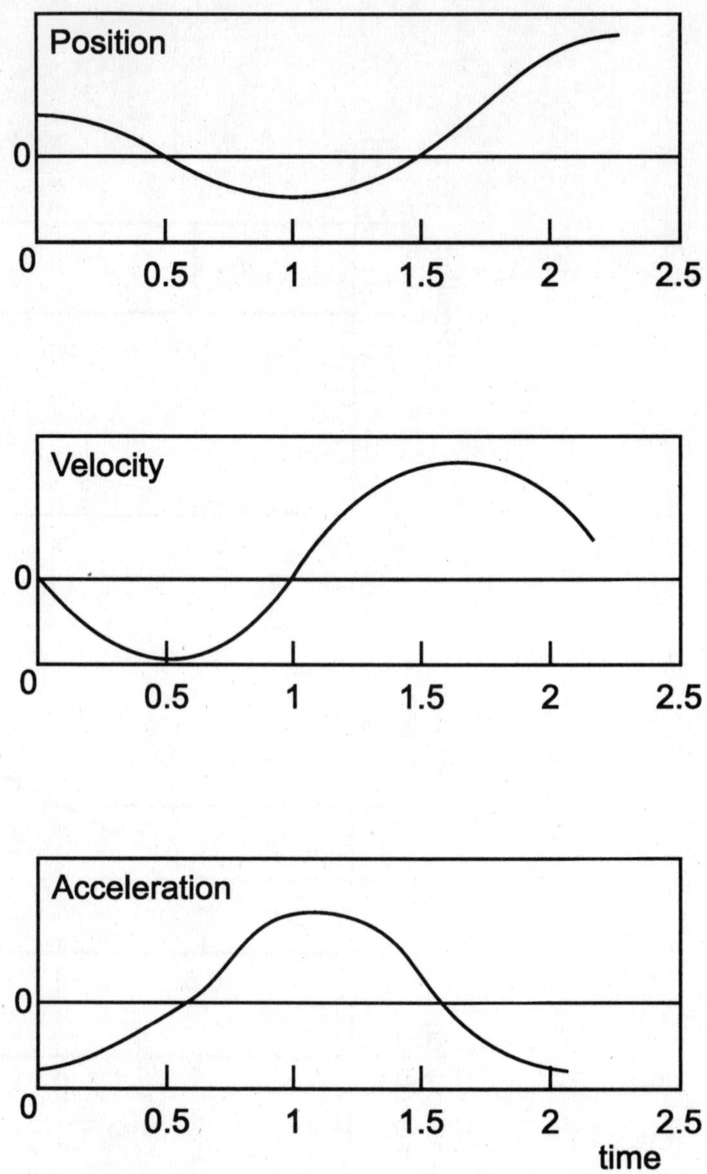

Calculus users, note that the formula used for the position in the graphs above is $x = 2\cos(3t) + t^2 - 1$.

MOTION IN TWO DIMENSIONS, INCLUDING PROJECTILE MOTION

COORDINATE SYSTEM AND VECTORS

Motion is a change of position, so to discuss motion, we must first discuss position. To discuss position, we must choose an origin and a reference direction. These choices are arbitrary and must be made before the position can be defined.

DISTANCE/ANGLE METHOD

As in one dimension, we may describe the position of an object by its distance from the origin and the direction in which it is displaced from the origin.

We choose a location for the origin and a reference direction. Traditionally, the reference direction points to the right along a horizontal straight line.

We draw an arrow from the origin to the object. The length of the arrow is a distance and is called the *magnitude* of the position *vector.*

For us, a *vector* is a quantity that has both magnitude and direction.

The angle that the arrow line makes with the reference direction is taken as the *direction* of the position *vector.*

COMPONENT METHOD

For this description, we add a second reference direction. Calling the original direction (traditionally to the right) the *x direction,* our second direction is called the *y direction.*

The *y* direction is by definition perpendicular to the *x* direction. Traditionally, this is taken to be upwards on the page. When *y* points up (instead of down), with *x* to the right, we say that we have a *right handed* coordinate system.

The component description of a vector tells how much of the vector is along the *x* direction and how much is along the *y* direction. The figures below show two different vectors and their components.

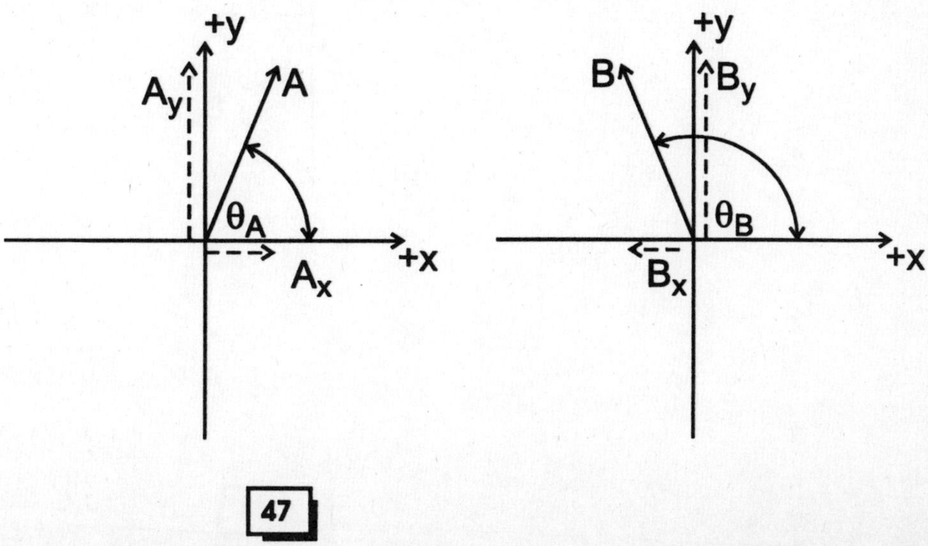

VECTOR ALGEBRA

Vector Addition and Subtraction—Vectors are added following rules that work for adding steps in a journey on foot. $\vec{A}+\vec{B}$ is determined by placing the tail of the $\vec{B}$ vector on the tip of the $\vec{A}$ vector. The resultant vector, $\vec{A}+\vec{B}$ is the vector from the tail of $\vec{A}$ to the tip of $\vec{B}$, as in the figure below.

The vector difference $\vec{A}-\vec{B}$ is as the sum of the vector $\vec{A}$, with the "reverse" of the vector $\vec{B}$, called $-\vec{B}$. As in the figure, $-\vec{B}$ points opposite to $\vec{B}$.

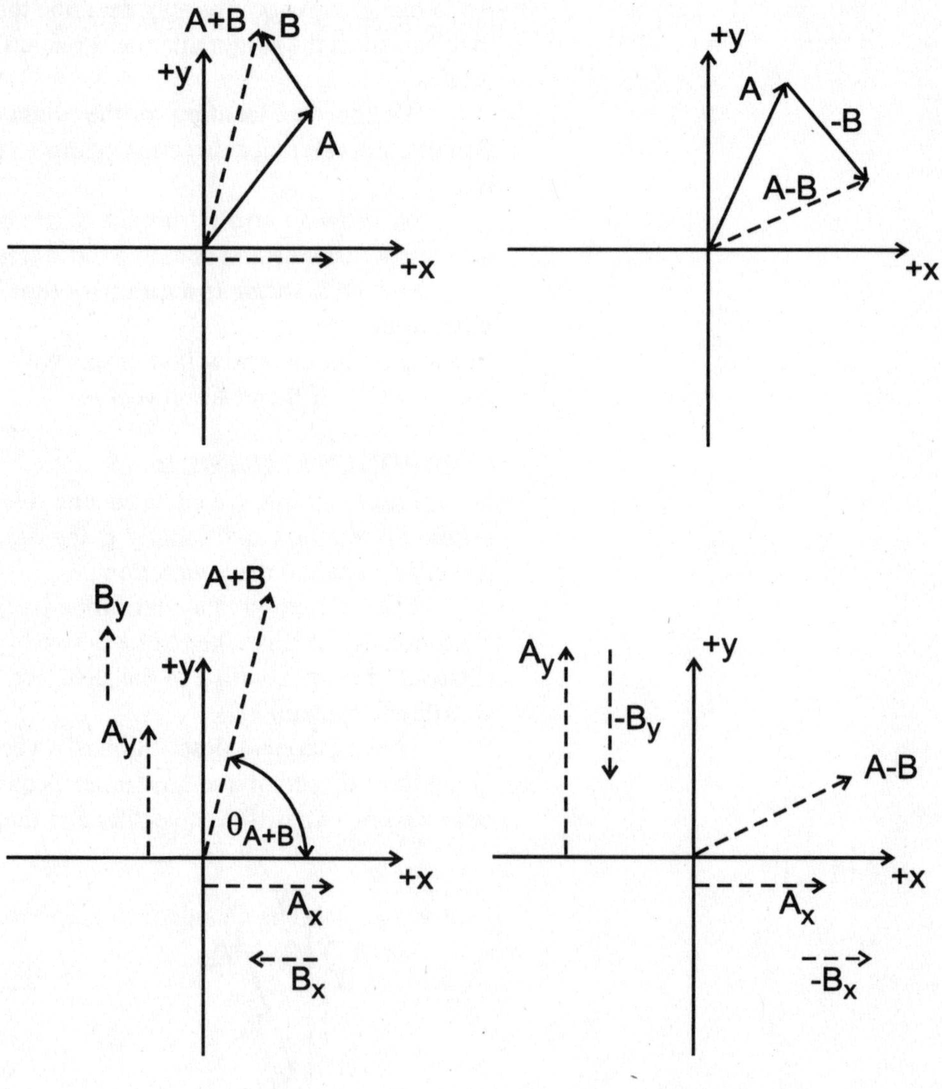

Using the component method as sketched above, the magnitude of the vector sum of $\vec{A}$ and $\vec{B}$ is

$$\left|\vec{A}+\vec{B}\right| = \sqrt{[A_x+B_x]^2+[A_y+B_y]^2} = \sqrt{[\vec{A}+\vec{B}]_x{}^2+[\vec{A}+\vec{B}]_y{}^2}\ .$$

The angle, θ, between $\vec{A}+\vec{B}$ and the reference direction (x=axis) is given by

$$\tan\theta_{\vec{A}+\vec{B}} = \frac{[A_y+B_y]}{[A_x+B_x]} = \frac{[\vec{A}+\vec{B}]_y}{[\vec{A}+\vec{B}]_x}\ .$$

The results for the difference between $\vec{A}$ and $\vec{B}$ can be found by simply placing a – sign in front of each $\vec{B}$ above.

Scalar Multiplication of a Vector—When a vector is multiplied by a scalar (a number with no direction) the magnitude (length) of the vector is multiplied by that number, and the direction is unchanged.

Scalar ("dot") Product of Two Vectors—The scalar product, $\vec{A}\bullet\vec{B} = AB\cos(\theta) = [A_xB_x]+[A_yB_y]$ is the product of the magnitudes times the cosine of the angle between the two vectors. This product can be positive or negative, depending on the sign of the cosine.

Vector ("cross") Product of Two Vectors—The vector product of two vectors has magnitude $/\ \vec{A}\times\vec{B}\ / = AB\sin(\theta)$.

The vector product also has direction. It is perpendicular to the plane defined by the $\vec{A}$ and $\vec{B}$ vectors. There are two directions that satisfy that condition. The ambiguity is resolved with a right-hand rule.

If $\vec{A}$ and $\vec{B}$ lie in the page, and if $\vec{A}$ must turn clockwise to become parallel to $\vec{B}$, then the direction of $\vec{A}\times\vec{B}$ is into the page.

Note that $\vec{A}\times\vec{B} = -\vec{B}\times\vec{A}$

DEFINITIONS

Displacement—The change in position is calculated by subtracting the initial position vector from the final position vector.

The algebraic representation of this calculation uses $\vec{r}$ as the symbol for position and $\Delta\vec{r}$ as the symbol for change in position. We write:

$$\Delta\vec{r} = \vec{r}_{FINAL} - \vec{r}_{INITIAL}$$

The arrows above $\vec{r}$ emphasize the vector nature of the position by reminding us that it has a direction as well as a magnitude.

$\Delta\vec{r}$ is called the *displacement* of the object.

Velocity—Velocity is defined as the rate of change of position. Average velocity vector, $<\vec{v}>$, is defined as the change in position, $\Delta\vec{r}$, divided by the time required to make the change, Δt:

$$<\vec{v}> = \frac{\Delta\vec{r}}{\Delta t}$$

The instantaneous velocity vector is the limit of the average velocity as the time interval (and thus also the change in position) approaches zero. It is the time derivative of the position.

$$\vec{v} = \lim_{\Delta t \to 0} \frac{\Delta\vec{r}}{\Delta t} = \frac{d\vec{r}}{dt}$$

Acceleration—Acceleration is defined as the rate of change of velocity. Average acceleration, $<\vec{a}>$, is defined as the change in velocity, $\Delta\vec{v}$, divided by the time required to make the change, Δt:

$$<a> = \frac{\Delta v}{\Delta t}$$

Instantaneous acceleration is the limit of the average acceleration as the time interval (and thus also the change in velocity) approaches zero. It is the time derivative of the velocity.

$$\vec{a} = \lim_{\Delta t \to 0} \frac{\Delta\vec{v}}{\Delta t} = \frac{d\vec{v}}{dt}$$

PROJECTILE MOTION WITH CONSTANT ACCELERATION

THE EXPERIMENTAL SITUATION

Near the surface of the earth, it is found that all *freely falling* objects have the same acceleration, $\vec{a} = \left[9.8\frac{m}{sec^2}, DOWNWARD\right] \equiv \vec{g}$.

Freely falling means that no force except gravity acts on the object. In particular, we ignore wind resistance. When an object is launched into the air with some initial velocity, it is freely falling after launch, even though it might not be moving downward.

The motion of such an object near the surface of the earth is the simplest example of projectile motion.

THE EQUATIONS OF MOTION AS A FUNCTION OF TIME

If we choose our *x*-axis to be horizontal and our *y*-axis to be vertical (the traditional choices) then the acceleration in the *x* direction is zero, while the acceleration in the *y* direction is $-9.8\frac{m}{sec^2}$.

In these coordinates, the projectile motion separates; the *x* motion is independent of the *y* motion (as long as the object is free falling). The equation of motion for the *y* component of the motion is

$$y = \frac{1}{2}\left(-9.8\frac{m}{sec^2}\right)t^2 + v_{0y}t + y_0$$

where y_0 is the *y* component of the initial position, and v_{0y} is the *y* component of the initial velocity.

(The *x* and *y* components of initial velocity are found from the magnitude and direction of the initial velocity. One uses the same trigonometric method as was used to find the *x* and *y* components of a position vector.) The *y* component of the velocity can be obtained by differentiation:

$$v_y = \left(-9.8\frac{m}{sec^2}\right)t + v_{0y}$$

The *x* component of the motion is more simple, since there is no acceleration in the *x* direction:

$$x = v_{0x}t + x_0$$

and the *x* component of the velocity is constant:

$$v_x = v_{0x}.$$

THE TRAJECTORY

The trajectory is a plot of the *y* component of the motion versus the *x* component of the motion. Each point on the trajectory represents the position at a particular time. It can be found by solving the *x* equation for *t* and replacing *t* in the *y* equation:

$$y = \frac{1}{2}\left(-9.8\frac{m}{sec^2}\right)\left[\frac{x - x_0}{v_{0x}}\right]^2 + v_{0y}\left[\frac{x - x_0}{v_{0x}}\right] + y_0$$

This simplifies if we choose our origin so that $x_0 = 0$ and $y_0 = 0$:

$$y = \left(-4.9\frac{m}{sec^2}\right)\left[\frac{x}{v_{0x}}\right]^2 + v_{0y}\left[\frac{x}{v_{0x}}\right]$$

In either case, the trajectory is a segment of a parabola, curving downward. We may ask, for example, for the *x* component of the position when the particle has returned to its initial height. The answer is often called the range of the projectile.

CIRCULAR MOTION

TWO DIMENSIONAL VECTOR DESCRIPTION

For an object moving in a circular path, we choose the origin of the coordinate system to be at the center of the circle. The position vector has constant magnitude r and a direction that varies as the object moves around the circle.

The velocity vector, $\vec{v} = \frac{d\vec{r}}{dt}$, is non-zero solely because of the changing direction of $\vec{r}$. The velocity vector is tangent to the circular path and always perpendicular to $\vec{r}$.

The acceleration vector, $\vec{a} = \frac{d\vec{v}}{dt}$, has two components—a radial component and a tangential component.

The radial component is called the centripetal acceleration. It is directed towards the center of the circle and has magnitude $|\vec{a}_C| = \frac{v^2}{r}$.

The tangential component of acceleration is zero if the magnitude of the velocity is constant. Otherwise, it causes changes in the velocity magnitude.

If the velocity is constant, $|\vec{v}| = \frac{2\pi r}{T}$,where T is the period of the motion—the time to make one round trip. For constant velocity,

$|\vec{a}_C| = \frac{4\pi^2 r}{T^2}$, and the tangential acceleration is zero.

REDUCTION TO A ONE DIMENSIONAL DESCRIPTION

If we narrow down our focus to the circular path, like a driver on a circular race track, circular motion can be described as one dimensional motion (on a curved path).

Position is measured from a reference position on the curve (traditionally the intersection of the x-axis with the curve, with the x-axis passing through the center of the circle). Traditionally, counter-clockwise is taken to be the + direction around the circle.

We name the position along the circle s. s is a one-dimensional vector.

The velocity is also one-dimensional: $v = \frac{ds}{dt}$. v is tangent to the circle.

When we use this one dimensional model, we ignore the centripetal acceleration. The only acceleration is the component tangent to the circle,

$a = \frac{dv}{dt}$. If the object moves at constant speed, $v = \text{constant}$, the one-dimensional acceleration (tangential acceleration) is zero.

ANGULAR DESCRIPTION

Continue to use an x-axis that passes through the center of the circle as a reference. The position vector, $\vec{r}$, makes an angle θ with the x-axis. Traditionally, θ is positive when it opens in the counterclockwise direction.

Taking the magnitude of $\vec{r}$ as given, θ measures the *angular position* of the object.

We define angular velocity, $\omega = \dfrac{d\theta}{dt}$, and angular acceleration, $\alpha = \dfrac{d\omega}{dt}$. For constant angular acceleration, we have the usual equations for one dimensional motion with constant acceleration:

$\theta = \dfrac{1}{2}\alpha t^2 + \omega_0 t + \theta_0$ and $\omega = \alpha t + \omega_0$, where the initial angular velocity and initial angular position are ω_0 and θ_0, respectively.

The relation between the angular and the linear description follows:

$$s = r\theta$$

$$|\vec{v}| = v = r\omega$$

$$a_{TANGENTIAL} = r\alpha$$

$$a_{CENTRIPETAL} = r\omega^2$$

TWO DIMENSIONAL COMPONENT DESCRIPTION

Taking the usual x- and y-axes—x horizontal and y vertical, with the origin at the center of the circle, and using θ as defined above, we can write the vector components of the position of an object that follows a circular path:

$$x = r\cos\theta \text{ and } y = r\sin\theta$$

We limit ourselves to the special case of zero angular acceleration. Then,

$$\theta = \omega_0 t + \theta_0 \text{ and } \omega = \omega_0 \text{ so } \theta = \omega t + \theta_0 \cdot$$

We have $x = r\cos(\omega t + \theta_0)$ and $y = r\sin(\omega t + \theta_0)$. Then the velocity components are $v_x = -\omega r\sin(\omega t + \theta_0)$ and $v_y = \omega r\cos(\omega t + \theta_0)$. The acceleration components are

$$a_x = -\omega^2 r\cos(\omega t + \theta_0) \text{ and } a_y = -\omega^2 r\sin(\omega t + \theta_0) \cdot$$

Chapter 2
NEWTON'S LAWS OF MOTION

EQUILIBRIUM (FIRST LAW)

If you give an object a position and then arrange things so that it is "left alone," it keeps the position that you gave it.

If you give an object a velocity and then arrange things so that it is left alone, it keeps the velocity (both magnitude and direction) that you gave it.

If you give an object an acceleration and then arrange things so that it is left alone, the acceleration drops to zero the moment that you release it. **Sir Isaac Newton wrote, "Every body continues in its state of rest, or of uniform motion in a right [straight] line, unless it is compelled to change that state by forces impressed upon it."**— *Principia, Motte's 1729 translation into English, revised by Cajori, University of California Press, 1934, p.13.*

DYNAMICS OF A SINGLE PARTICLE (SECOND LAW)

The second and third laws deal with the way that forces change the velocity of an object. A standard form for the second law is $\vec{F}_{TOTAL} = m\vec{a}$, where $\vec{a}$ is the acceleration vector, m is the mass of the object, and $\vec{F}_{TOTAL}$ is the vector sum of all the forces applied *to* the object. The news in the second law is not that $m\vec{a}$ is a force—that information is available in the third law. The news is in the word TOTAL, and what it means.

Forces add like vectors, and the total force is calculated by adding up *all* of the forces *on* the object, using the rules developed for adding displacement vectors. Newton wrote the law in terms of momentum: $\vec{F}_{TOTAL} = \frac{d\vec{p}}{dt}$, where $\vec{p} = m\vec{v}$ is the momentum.

AN EXAMPLE FOR THE SECOND LAW

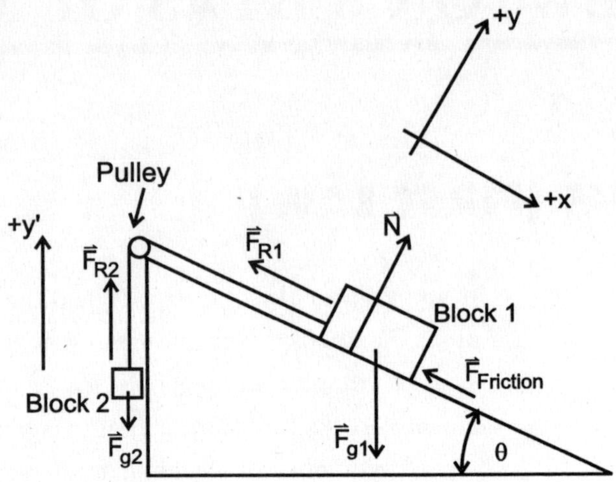

The figure shows most of the forces that appear in mechanics problems. The second law can be applied separately to the two blocks. The magnitude of the rope force is its tension, $\left|\vec{F}_{R2}\right| = T_2$ and $\left|\vec{F}_{R1}\right| = T_1$.

If the pulley is ideal, the two tensions are equal. If not, the difference in the tensions provides the force necessary to move the pulley.

For block 2, the y' component of the second law is $T_2 - m_2 g = m_2 a_{2y'}$ where m_2 is the mass of block 2.

For block 1, there are more forces. Again, there is a force by the rope and the force of gravity, although now those two do not act along the same line. In addition, there are forces by the ramp. In the figure, two forces by the ramp on block 1 are named.

$\vec{N}$ is the normal force, acting perpendicular to the ramp surface, and $\vec{F}_{FRICTION}$ is the friction force, acting parallel to the ramp surface. In mechanics problems the *magnitudes* of those two forces are related in a fairly simple way: $\left|\vec{F}_{FRICTION}\right| = \mu_k \left|\vec{N}\right|$ where μ_k is called the coefficient of kinetic (or sliding) friction.

The direction of the friction force is always opposite to the direction of the sliding motion. In the figure, it is assumed that block 1 is sliding down the ramp.

If the block is not sliding, the relation is more complicated. The static friction force adjusts to whatever value is necessary to keep the block at rest. There is an upper limit to how much static friction force the ramp can apply to the block. It is calculated using $\left|\vec{F}^{MAXIMUM}_{STATIC\ FRICTION}\right| = \mu_s\left|\vec{N}\right|$ where μ_s is called the coefficient of static friction.

The two components (note that block 1 is using a different, tilted, coordinate system) for the second law in the case of block 1 are

- $N - m_1 g\cos\theta + 0 = m_1 a_{1y} = 0$ for the y component, since the rope is parallel to the block surface and there is no acceleration of the block up off the ramp.

- $-\mu_k m_1 g\cos\theta + m_1 g\sin\theta - T_1 = m_1 a_{1x}$ for the x component, assuming that the block is sliding downhill.

When solving problems like this, it is useful to note that, because the rope does not stretch or break, the accelerations of the two blocks have the same magnitude, even though their directions are different.

TWO FORCES

Near the surface of the earth, the force of gravity on an object of mass m is $\vec{F}_g = m\vec{g}$ where $\vec{g}$ is the gravitational field strength. The magnitude of $\vec{g}$ is $9.8m/\sec^2$ (the acceleration of a freely falling object) and the direction of $\vec{g}$ is downward.

When a spring is displaced by a distance x from its equilibrium (relaxed) position, it exerts a force in the x direction, $F_{SPRING} = -kx$. k is the spring constant for the spring. The force always acts to return the spring to its equilibrium position.

Chapter 3

WORK, ENERGY, POWER

WORK AND WORK-ENERGY THEOREM

DEFINITIONS OF WORK

Unless otherwise stated, work is done *on* an object by a force (or combination of forces). In some cases, it is useful to study the work done *by* a particular force (among other forces) acting on an object, but work *on* the object will be our primary concern.

- For a single constant force acting parallel to the (straight-line) motion of an object, the work done on the object is equal to the product of the magnitude of the force and the distance traveled. If the direction of the force is antiparallel (opposite) to the direction of the displacement, the work is negative.

- The units of work are Newton-meters. Work is given its own unit, the Joule. $1\ Joule = 1\ Newton\text{ - }meter$

- For a constant force $\vec{F}$ acting on an object that moves in a straight line through a displacement, $\vec{s}$, the work is the product $W = |\vec{F}||\Delta\vec{s}|\cos\theta$, where θ is the angle between the two arrows representing the force and the displacement vectors. (We use *s* rather than *r* to suggest that the path might be curved in other problems.) If the cosine function is negative for the angle in the problem, the work is negative.

- For a force $\vec{F}$ (not necessarily constant) acting on an object that follows a path (possibly curved) *S*, the work by the force on the object is the integral

$$W = \int_A^B |\vec{F}||d\vec{s}|\cos\theta = \int_A^B \vec{F} \bullet d\vec{s} = \int_A^B [F_x dx + F_y dy + F_z dz]$$

The integral is a path integral along path S, from point A on the path to point B on the path. $d\vec{s}$ is an increment along the path, which can be expressed in its x, y, and z components.

In general, the work done when the object moves from point A to point B depends on the path, S, that is followed. A different path may be expected to produce different work, even though the beginning and end are the same.

The Work-Energy Theorem

Now, we take the force in the definition of work to be the **total** force on the object. In that case, we can replace the total force using Newton's second law. The work can be calculated in general. The result is

$$\begin{aligned} W_{NET}^{ON\ OBJECT} &= \left[\frac{1}{2}mv_{FINAL}^2\right] - \left[\frac{1}{2}mv_{INITIAL}^2\right] \\ &= KE_{FINAL} - KE_{INITIAL} \\ &= \Delta KE \end{aligned}$$

$KE = \frac{1}{2}mv^2$ is called the kinetic energy of the object.

CONSERVATIVE FORCES AND POTENTIAL ENERGY

- The formal meaning of conservative forces is if the work done by a particular force is independent of path, then the work done to move an object from A to B is simply a function of the positions, $\vec{r}_A$ and $\vec{r}_B$. (In calculus terms, the integral is analytic. It may be done once and for all using the most convenient path for calculation. The integral is independent of the path.)

This happens when the force is a function of position only. (The most common such force is the force of gravity. Near the surface of the earth, it is a constant function. Even at great distance from the earth, the gravitational force on an object is a simple function of the distance to the center of the earth.)

If we pick a reference point, agreeing to always take A to be at that point, the work to get from the reference point to a point $\vec{r}$ is simply a function of $\vec{r}$.

- When an object moves from point A to point B, the work done by a conservative force is exactly the negative of the work it does when the object moves from point B to point A.

- **A potential energy example**—If we lift a box from the floor (point *A*) up to a shelf (point *B*), the force of gravity does negative work in the process. If the box falls from the shelf to the floor (from *B* to *A*) the force of gravity does positive work equal in magnitude to the negative work that it did when the box was on the way up.

The negative work done on the way up is a measure of the ability of gravity to do work on the box should it fall. We call that ability to do work *potential energy* and give it the symbol *U.*

- For a conservative force, the work done by the force is the negative of the change in the potential energy associated with the force.

For gravity near the surface of the earth, $U = mgh$. This is the equation for the potential energy of an object of mass *m*, located at a height h above the reference location.

For a spring, whose restoring force $\vec{F}$ is related to its displacement from equilibrium, x, by $\vec{F} = -kx$, the potential energy is $U_{SPRING} = \frac{1}{2}kx^2$

In general, for a conservative force, $\vec{F}$, $\Delta U_{AB} = U_B - U_A = -\int_A^B \vec{F} \bullet d\vec{s}$.

Conversely, the force may be calculated from the potential energy function by differentiation. The *x* component of the force is

$F_x = -\frac{\partial U}{\partial x}$, with similar expressions for the *y* and *z* components of the force. The derivative shown is a partial derivative, meaning that *y* and *z* are held constant while it is calculated.

CONSERVATION OF ENERGY

The work-energy theorem can be rewritten with each force that goes into the total force written separately. The work done by the conservative forces is the negative of their potential energy changes. After rearranging, the work energy theorem takes the form $\Delta U + \Delta KE = W_{\text{OTHER FORCES}}$. There may be more than one potential energy. In that case, each is represented by a ΔU on the left side. The other forces are the nonconservative forces in the problem. The most common other force is the force of friction.

If there are no nonconservative forces in the situation, then energy is said to be conserved. We write

$$\Delta U + \Delta KE = 0 \text{ or}$$

$$U_{FINAL} + KE_{FINAL} = U_{INITIAL} + KE_{INITIAL} \text{ or}$$

$$E_{FINAL} = E + E_{INITIAL} \quad \text{where } E = U + KE$$

is called the total energy of the object.

POWER

Power is the rate of doing work. Power has the units $Joule / sec = Watt$
Power can be calculated in two ways:

$$P = \frac{dW}{dt} \text{ or}$$

$$P = \vec{F} \bullet \vec{v}$$

Positive power means that energy is being added to the object.

Chapter 4

SYSTEMS OF PARTICLES, LINEAR MOMENTUM

CENTER OF MASS

The location of the center of mass of a system of N particles is calculated

from $\vec{R}_{CM} = \frac{\sum_{i=1}^{N} \vec{r}_i m_i}{\sum_{i=1}^{N} m_i}$, where the subscript i refers to the "ith" particle in

the system. The numerator is the sum of all the products $\vec{r}_i m_i$, one for each object. The denominator, the sum of all the masses, is the total mass of the system.

For an extended object of density ρ, the sum becomes an integral:

$$\vec{R}_{CM} = \frac{\int_{VOLUME} \vec{r} \rho dV}{\int_{VOLUME} \rho dV}$$ where dV is an element of volume.

For uniform objects, the center of mass is at the geometric center of the object.

When forces are applied to a uniform object, such as a baseball, Newton's second law correctly calculates the acceleration of the center of mass, using the simple vector sum of the forces for the total force.

The velocity and position as functions of time (calculated from that acceleration) are the velocity and position of the center of mass of the object.

IMPULSE AND MOMENTUM

When an object is involved in a collision, such as a baseball colliding with a bat, we often know the initial and final velocity of the ball and would like to calculate the force on the ball. If the time duration of the collision is Δt, we can calculate the average force from

$$\langle \vec{F} \rangle = m\langle \vec{a} \rangle = m\frac{\Delta \vec{v}}{\Delta t} = m\frac{\vec{v}_{FINAL} - \vec{v}_{INITIAL}}{\Delta t}.$$

We often do not know the time interval. We can combine the things that we do know, writing $\langle \vec{F} \rangle = \frac{m\vec{v}_{FINAL} - m\vec{v}_{INITIAL}}{\Delta t} = \frac{\Delta \vec{p}}{\Delta t}$, where we have defined the *momentum*, $\vec{p} = m\vec{v}$. (Newton actually used momentum in his version of the second law, written in modern form as $\vec{F}_{TOTAL} = \frac{d\vec{p}}{dt}$.)

We call the change in momentum of an object involved in a collision the *impulse:* $IMPULSE = \Delta\vec{p}$. If the time duration of the collision is known, the average force can be calculated as the impulse divided by the duration.

If the duration is known, and the forces are known as a function of time, the impulse may be calculated as the integral,

$$\int_{START\ TIME}^{END\ TIME} \vec{F}_{TOTAL}\, dt = \vec{p}_{FINAL} - \vec{p}_{INITIAL} = IMPULSE \cdot$$

CONSERVATION OF LINEAR MOMENTUM IN COLLISIONS: $\vec{P}^{BEFORE}_{SYSTEM} = \vec{P}^{AFTER}_{SYSTEM}$

If the total force on the object is zero, the final momentum is equal to the initial momentum, in agreement with Newton's first law. In this case, momentum is said to be conserved.

If two objects interact, they exert forces on each other, and we can expect the momentum of each object to change. There is an *object* in such a collision for which momentum is in fact conserved. This new, larger *object* is the *system* composed of the two objects together.

We can define the momentum of the system as the sum of the individual momenta. In general, $\vec{P}_{SYSTEM} = \sum_{i=1}^{N} m_i \vec{v}_i$ for a system of *N* particles. If the collection of objects experiences no outside forces, then the momentum of the system remains constant, $\vec{P}^{BEFORE}_{SYSTEM} = \vec{P}^{AFTER}_{SYSTEM}$, even though the momentum of some of the individual members of the system might change. Note that the momentum of the system is simply related to the velocity of the center of mass of the system: $\vec{P}_{SYSTEM} = m_{SYSTEM}\vec{v}_{CENTER\ OF\ MASS}$.

If momentum is conserved for the system, then the velocity of the center of mass remains constant, even though pieces of the system might go flying off in new directions.

Since the momentum is a vector, constant momentum means that each component of the momentum is constant, separately. Depending on the problem, this can mean that conservation of momentum during a collision gives two or even three equations to work with while solving the problem. One question must be asked before applying conservation of momentum, "Is the net force on the system zero?" Unless the answer is yes, the law of conservation of momentum cannot be applied to that system. Often the answer can be changed from no to yes by increasing the number of objects in the system.

COLLISIONS THAT CONSERVE BOTH ENERGY AND MOMENTUM

If energy and momentum are both conserved in a collision, the collision is said to be *elastic*. If energy is not conserved, the collision is *inelastic*.

One-Dimensional Case

In one-dimensional elastic collisions between two objects, an explicit pair of equations can be written to relate the initial velocities to the final velocities. The equations are

$$v_1' = \left[\frac{m_1 - m_2}{m_1 + m_2}\right] v_1 + \left[\frac{2m_2}{m_1 + m_2}\right] v_2 \text{ and}$$

$$v_2' = \left[\frac{2m_1}{m_1 + m_2}\right] v_1 + \left[\frac{m_2 - m_1}{m_1 + m_2}\right] v_2$$

In these equations, the subscript 1 is attached to the mass and velocities of object 1, and 2 is attached to the mass and velocities of object 2. The ′ mark on a velocity indicates a velocity after the collision. Unmarked velocities are velocities observed before the collision.

Without loss of generality, we may imagine that the plus direction is to the right, and that object 2 is initially to the right of object 1. The equations are useful in examples such as a ball bouncing from a wall or from a tennis racquet.

Two-Dimensional Case

In two dimensions, there are not enough equations to solve for the final velocities in terms of the initial velocities. The conservation laws can be written in terms of momentum. We show the special case in which both objects have the same mass and object 2 is initially at rest.

$$\vec{p}_1' + \vec{p}_2' = \vec{p}_1$$

$$\left|\vec{p}_1'\right|^2 + \left|\vec{p}_2'\right|^2 = \left|\vec{p}_1\right|^2.$$

The upper equation comes from conservation of momentum. The lower equation is conservation of energy, assuming that there is no change in the potential energy of the objects. The standard example of this collision is a collision between billiard balls.These two equations show that the angle between $\vec{p}_1^{\,'}$ and $\vec{p}_2^{\,'}$ must be 90 degrees.

Chapter 5

CIRCULAR MOTION AND ROTATION

UNIFORM CIRCULAR MOTION

Uniform circular motion is covered in Chapter 1.

ANGULAR MOMENTUM AND ITS CONSERVATION

POINT PARTICLES

We first define angular momentum for a point object moving on a circle of radius $|\vec{r}|$. Taking the origin to be at the center of the circle, the position vector of the object is $\vec{r}$. For circular motion, the velocity vector, $\vec{v}$, is perpendicular to $\vec{r}$ and so is the momentum vector, $\vec{p} = m\vec{v}$.

The angular momentum vector, $\vec{L}$, is the vector product of the radius vector with the momentum vector: $\vec{L} = \vec{r} \times \vec{p}$. Traditionally, the angular momentum is positive when the rotation is counterclockwise. The vector product is discussed in Chapter 1.

This definition works with all motion—not just circular motion. Pick an origin, and the angular momentum of a particle about that origin may be calculated, no matter what the shape of its path.

EXTENDED SYSTEMS

For a system of several particles, the angular momentum of the system is the sum of the angular momenta of each particle.

RIGID EXTENDED BODIES

A rigid object may be treated as if it is composed of many small parts. The angular momentum of the rigid body is the sum of the angular momenta of each of its parts.

For a rotating rigid body, every point in the body has the same angular velocity, ω. The tangential velocity of a point at position $\vec{r}$ within the body is $v = |\vec{r}|\omega = r\omega$. The momentum of each bit of the object can be calculated and added to find the angular momentum of the rigid object.

Factoring out the constant angular velocity gives $|\vec{L}| = L = I\omega$, where I represents the geometrical arrangement of the mass within the rigid body. I is called the *moment of inertia* for the object. If we ignore the spokes and the thickness of the rim, the moment of inertia for a spinning bicycle wheel is $I = mR^2$ where m is the mass of the wheel and R is its radius. (It is assumed that the wheel is spinning in the usual way on its axle.)

ROTATIONAL INERTIA

ABOUT CENTER OF MASS

For a rigid body, the moment of inertia about an axis of rotation through the center of mass is given by an integral over the volume of the object,

$$I_{CM} = \int_{VOLUME} r^2 \rho dV .$$

The density, ρ, is generally a function of position, but in most problems is a constant. r is the magnitude of the perpendicular distance from the axis of rotation to the volume element dV.

Ignoring the hole in the center, the moment of inertia (about the center of mass) of a compact disk as it is spun in its drive is $I_{CM} = \frac{1}{2}mR^2$. R is the radius of the disk and m is its mass.

ABOUT ANOTHER AXIS: THE PARALLEL AXIS THEOREM

If the moment of inertia for an object rotating about a particular axis through the center of mass is I_{CM}, the moment of inertia about any axis parallel to the original axis is $I = I_{CM} + mD^2$, where D is the perpendicular distance between the two axes. For a rolling compact disc, the axis of rotation is at a point on the rim. The moment of inertia about this axis is

$$I = \frac{1}{2}mR^2 + mR^2 = \frac{3}{2}mR^2 .$$

PARTICLE MOVING IN A CIRCLE

The parallel axis theorem shows that the moment of inertia of a small object revolving on the end of a long string is $I = mR^2$, where R is the length of the string (and the radius of the circular path).

CONSERVATION OF ANGULAR MOMENTUM

If an object rotates with no external influences, its angular momentum remains constant. In this situation, we say that angular momentum is conserved.

If the object changes its moment of inertia by changing its geometrical shape and there are still no external influences, then angular momentum is still conserved.

If the object changes its moment of inertia while angular momentum is being conserved, then the angular velocity must change to keep the angular momentum constant:

$L = I_{BEFORE}\omega_{BEFORE} = I_{AFTER}\omega_{AFTER}$.

Skaters use this trick to spin faster by pulling in their arms and legs, thereby reducing their moment of inertia.

TORQUE AND ROTATIONAL STATICS

DEFINITION OF TORQUE

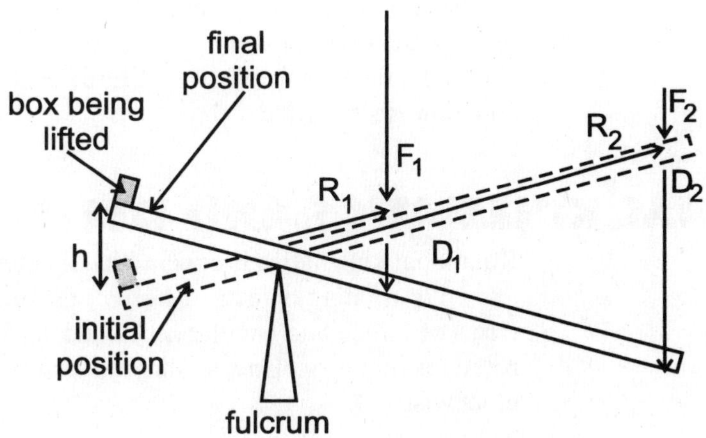

The figure shows a box being lifted through height *h* by a lever. Two different methods are shown. In one case a large force $\vec{F}_1$ is applied through a small distance D_1. In the second case, a small force $\vec{F}_2$ is applied through a large distance D_2.

For a massless lever, the work done in each case is *mgh*, the change in gravitational potential energy of the box of mass *m*. Thus the products F_1D_1 and F_2D_2 must be equal. A little geometry shows that the products R_1F_1 and R_2F_2 must also be equal.

Torque is defined so as to represent this result:

$\tau = rF\sin\theta$, where θ is the angle between the force vector and the vector from the axis of rotation to the point of application of the force. When the angle is 90 degrees, we have the simple expression $\tau = rF$.

Traditionally, a torque that tends to produce counterclockwise motion is taken as a positive torque. The two forces in the figure produce negative torques.

Formally, the torque vector is defined as the vector product

$$\vec{\tau} = \vec{r} \times \vec{F}$$

STATIC EQUILIBRIUM

The second static object must obey two conditions:

1. The total force on the object is zero.
2. The total torque on the object is zero.

This condition can also be stated as *"the total clockwise torque must equal the total counterclockwise torque."*

In static equilibrium, the center of rotation for calculation of torque can be chosen for convenience. Typically, you choose the center of rotation through the point of application of one of the unknown forces on the object. That way, the torque due to that force is zero, and the torque equilibrium equation is simplified.

ROTATIONAL KINEMATICS AND DYNAMICS

Rotational kinematics is covered in Chapter 1.

For rotating objects, the effect of force on acceleration is rewritten in terms of torque and angular acceleration. We restrict consideration to rotations in the *xy* plane, so that the torques are clockwise (–) or counterclockwise (+).

$\tau_{TOTAL} = I\alpha$ where τ_{TOTAL} is the sum of all the torques applied to the object, + for counterclockwise and – for clockwise torques. α is the angular acceleration, and I is the moment of inertia of the object.

The moment of inertia must be calculated for the same axis as the axis used for the torque. Because the object is rotating, the physical axis of rotation must be used; no other can be selected.

In terms of angular momentum, $\vec{L}$, $\vec{\tau}_{TOTAL} = \dfrac{d\vec{L}}{dt}$

Work is $W = \displaystyle\int_{\theta 1}^{\theta 2} \tau d\theta$

Kinetic energy is $KE_{ROTATION} = \dfrac{1}{2}I\omega^2$

Combined Rotation and Linear Motion

- When the object has both rotational and translational motion (such as a spinning football flying through the air),

  $KE = \frac{1}{2}mv^2 + \frac{1}{2}I\omega^2$ where v is the velocity of the center of mass of the object.

- When the object is *rolling*, the angular velocity about the axis of rotation is related to the center of mass velocity by $v = R\omega$, where R is the radius of the rolling object.

In that special case, $KE = \frac{1}{2}\left[m + \frac{I_{cm}}{R^2}\right]v^2$

Chapter 6

OSCILLATIONS AND GRAVITATION

SIMPLE HARMONIC MOTION (DYNAMICS AND ENERGY RELATIONSHIPS)

If Newton's second law for an object results in an equation of the form $ma = -kx$, then the solution is known mathematically:

$x = A\cos(\omega t + \phi)$, where A is the *amplitude* and $\omega = \frac{2\pi}{T} = 2\pi f$ is the angular frequency.

In order for the solution to work, it *must* be true that

$$\omega = \sqrt{\frac{k}{m}}.$$

T is the period of the motion, and f is the frequency of the motion. ϕ is a "phase angle" that is chosen to match initial conditions. It is often zero. If the cosine function is replaced by a sine function, the new expression is also a solution.

Velocity and acceleration in this solution are: $v = -\omega A\sin(\omega t + \phi)$ and $a = -\omega^2 A\cos(\omega t + \phi)$ for the cosine solution and $v = \omega A\cos(\omega t + \phi)$ and $a = -\omega^2 A\sin(\omega t + \phi)$ for the sine solution.

It is important to note that $a = -\omega^2 x$. This is the signature of simple harmonic motion that is true for all forms of the solutions.

The kinetic energy is $KE = \frac{1}{2}K^2A^2\cos^2(\omega t + \phi)$. The maximum value of the kinetic energy is $KE_{MAXIMUM} = \frac{1}{2}K^2A^2$.

The potential energy associated with this motion is taken to be zero when the KE is maximum. Then, the total energy is $E = \frac{1}{2}K^2A^2$. Using conservation of energy and the Pythagorean theorem, we conclude that the potential energy is $U = \frac{1}{2}\omega^2A^2\sin^2(\omega t + \phi)$.

MASS ON A SPRING

The force exerted by a spring when it is displaced a distance, *x*, from its equilibrium position is $F_x = -kx$.

SPRING EXERTS THE ONLY FORCE

If a mass is attached to a horizontal spring and allowed to move on a frictionless horizontal surface, the only horizontal force on the mass is the spring force. We assume that the other end of the spring is fixed, so that the position of the mass, *x*, is also the displacement of the spring from its equilibrium length.

Then, Newton's second law for the horizontal direction is

$[F]_{TOTAL} = -kx = ma$, exactly that required for a simple harmonic motion solution. The angular frequency is $\omega = \sqrt{\frac{k}{m}}$. The frequency is

$f = \frac{1}{2\pi}\sqrt{\frac{k}{m}}$.

THE SPRING HANGS VERTICALLY SO THAT THE FORCE OF GRAVITY IS ALSO IMPORTANT

In the vertical direction, $[F]_{TOTAL} = -ky - mg = ma$, which is not immediately a simple harmonic motion equation. Adding a constant to *y* does not change its acceleration, so that we can replace *y* with $z = y + \frac{mg}{k}$. Then, the second law gives us

$[F]_{TOTAL} = -kz = ma$, and the solution is

$z = A\cos(\omega t + \phi)$, with $\omega = \sqrt{\frac{k}{m}}$.

This solution oscillates about $z = 0$, so in terms of *y*, the mass on the spring oscillates about $y_0 = -\frac{mg}{k}$. The only effect of the force of gravity is to shift the center of oscillation a distance $y_0 = -\frac{mg}{k}$.

PENDULUM AND OTHER OSCILLATIONS

A simple pendulum is a mass hung by a string from a point support. The mass is free to swing at the end of the string. The figure shows the definition of terms.

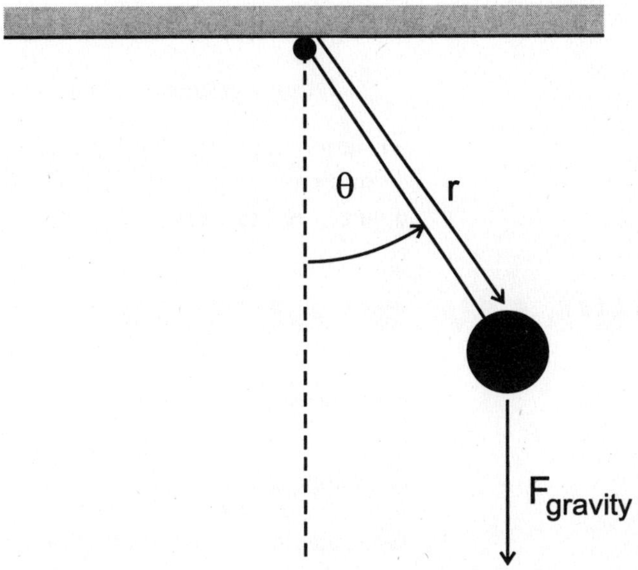

The angle θ is positive in the counterclockwise direction. The torque due the force of gravity is $\vec{\tau} = \vec{r} \times \vec{F}$ and in the picture is acting in the clockwise (–) direction. The magnitude of the torque is $\tau = rF\sin\theta$.

Newton's second law in rotational form is $\tau = I\alpha$ or

$-mgr\sin\theta = I\alpha$. This can be written as $\alpha = -\left[\frac{mgr}{I}\right]\sin\theta$, which is not quite the equation whose solution is simple harmonic motion. In order to make this into an equation that can be solved, we use the small angle approximation: When the angle is measured in radians, for small angles,

$\sin\theta \cong \theta$. In that case, we can write $\alpha = -\left[\frac{mgr}{I}\right]\theta$, with the immediate

solution $\theta = A\cos(\omega t + \phi)$ with $\omega = \sqrt{\frac{mgr}{I}}$.

One word of warning: In the equation above, ω is not the angular velocity of the pendulum. It is the angular frequency, $\omega = \frac{2\pi}{T}$.

In the case of the simple pendulum, $I = mr^2$, so the frequency is simply

$\omega = \sqrt{\frac{g}{r}}$, so that the period is $T = 2\pi\sqrt{\frac{r}{g}}$.

Other oscillators—Any system for which Newton's second law yields a relation between position and acceleration of the form $a = -[\ ]x$ is a simple harmonic oscillator. The angular frequency is the square root of the quantity in brackets.

NEWTON'S LAW OF GRAVITY

The gravitational force between two massive objects is always attractive. It has magnitude $F = G\frac{M_1 M_2}{r^2}$ where M_1 and M_2 are the masses of the two objects, r is the separation of their centers, and G is the universal gravitational constant, $G = 6.67\text{x}10^{-11}\,\frac{Newton\ m^2}{kg^2}$.

If there are no other forces on object M_2, its acceleration has magnitude $a = G\frac{M_1}{r^2}$ and is directed toward M_1. Near the surface of the earth, $g = a = G\frac{M_1}{r^2}$. Knowing the radius of the earth and G, this equation allows us to determine the mass of the earth.

ORBITS OF PLANETS AND SATELLITES

CIRCULAR

A planet orbiting the sun in a circular orbit has centripetal acceleration $a = \frac{v^2}{r}$. Assuming that the gravitational force is the only force acting, Newton's second law yields $T^2 = \frac{4\pi}{GM_{SUN}} r^3$. This was originally derived experimentally as *Kepler's third law*.

If we measure the distance from earth to the sun, and know the period of our orbit, we can calculate the mass of the sun.
A satellite orbiting the earth is particularly useful if its orbit speed keeps it above one spot on the earth's equator—that is, it has a period of revolution equal to 1 day. Kepler's third law tells us that there is only one value of r that will give such a *geostationary* orbit.

GENERAL

KEPLER'S FIRST LAW
The planets orbit in elliptical paths with the sun at one focus of the ellipse. Newton extended this to unbound orbits. All orbits are conic sections: circles, ellipses, parabolas, and hyperbolas. Parabolas and hyperbolas are the unbound orbits. A visiting body on such a path goes by the sun only once.

KEPLER'S SECOND LAW
Planets sweep out equal areas in equal times. This is equivalent to the law of conservation of angular momentum for the planets. When the radius is short, the orbit velocity is large, keeping the angular momentum constant.

KEPLER'S THIRD LAW
The square of the period (of a bound orbit) is proportional to the cube of the semimajor axis of the orbit. (An elliptical orbit has two axes. The narrowest diameter of the ellipse is the minor axis and the largest diameter is the major axis. The semimajor axis is half the largest diameter.)

UNIT 2
Thermal Physics

Chapter 7
TEMPERATURE AND HEAT

Just as time is the quantity measured by a clock, temperature is the quantity measured by a thermometer. Just as we intend for a clock to measure how long something takes to happen, we intend a thermometer to measure how hot something is.

The MKS thermometer is calibrated in degrees Celsius ($°C$); $0\,°C$ at the temperature of ice water and $100\,°C$ at the temperature of boiling water. Temperature is given the symbol *T*. The conversion from Celsius to Fahrenheit is given by $T(°F) = 32\,°F + \frac{9}{5}T(°C)$.

MECHANICAL EQUIVALENT OF HEAT

To raise the temperature of an object, we *heat* it. There are two traditional ways to heat an object:

- **Thermally**—e.g. place the object in a flame. We measure the heat added in units of the amount of heat (e.g. how long in the flame) that it takes to raise 1 gram of water by $1\,°C$. This unit of heat is called a *calorie*. Heat transfer is given the symbol *Q*.

- **Mechanically**—e.g., rub the object on a spinning wheel and let friction *heat it up*. We measure the work required to raise the temperature by measuring the work done by friction. The units here are *Joules*. Work is given the symbol *W*.

If we raise the temperature of an object by, say, $1\,°C$ thermally, and then repeat the experiment using the mechanical method, we find that there is no difference in the heated object for the two methods. We cannot look at the object and determine which method was used to heat it.

The mechanical work done by friction is equivalent to the heat added thermally. Experimentally, 4.184 Joule = 1 calorie. This is the mechanical equivalent of heat.

SPECIFIC HEAT AND LATENT HEAT (INCLUDING CALORIMETRY)

SPECIFIC HEAT

Identical objects made from different materials require different amounts of heat to increase their temperature by $1\,°C$. If an object requires a large amount of heat to raise temperature by $1\,°C$, it is said to have a large *heat capacity.*

The heat capacity of an object is proportional to its mass, *m*. The heat capacity for 1 gram of a material is called the *specific heat* for that material. Specific heat is determined experimentally for each different material, and the information is available in tables of data.

Heat capacity is given the symbol *C*, and specific heat is given the symbol *c*.

The heat required to raise the temperature may be calculated from $Q = mc\Delta T = C\Delta T$.

The units of *c* above are either $\frac{calories}{g \quad °C}$ or $\frac{Joule}{g \quad °C}$, depending on the units desired for the heat, *Q*.

The specific heat may also be given using *kg* for the mass unit:

$\frac{calories}{kg \quad °C}$ or $\frac{Joule}{kg \quad °C}$.

The specific heat may also be given using moles for the measure of the amount of material. This is called the *molar specific heat*. The units are

$\frac{calories}{mole \quad °C}$ or $\frac{Joule}{mole \quad °C}$.

LATENT HEAT

When a substance reaches a phase transition temperature, the specific heat equation fails. (Two prototype phase transitions are the melting of ice and the boiling of water.)

In these phase transitions, heat is added without raising the temperature. The thermal energy is used to convert the low temperature phase (e.g., water) to the high temperature phase (e.g. steam).

The amount of heat to convert one gram of material to the high temperature phase is called the *latent heat, L*. The heat required to convert a mass, *m*, can be calculated from $Q = mL$.

The units are $\frac{calories}{g}$ or $\frac{Joule}{g}$. As above, the units can vary between calories and Joules and between *g*, *kg*, and moles.

When a material is cooled to its lower temperature phase, the amount of heat that must be removed to convert a gram of material to the low temperature phase is calculated with the same equation.

HEAT TRANSFER AND THERMAL EXPANSION

When two objects are able to exchange heat, heat energy will flow from the hotter to the colder object.

If the objects are connected by material such as a solid piece of wire, the mechanism of heat transfer is called *conduction.*

- No material moves in heat conduction, but heat energy is transferred.
- For a wire of length L and cross-sectional area A, the rate at which heat energy is transferred from the hot to the cold object by conduction is given by $H = \frac{dQ}{dt} = \frac{EA(\Delta T)}{L}$, where E is called the thermal conductivity of the material from which the wire is made.
- The rate of heat flow increases with thicker wire and decreases if the wire becomes longer.
- The rate of heat flow increases if the temperature difference, ΔT, between the two objects is increased.

If the objects are connected by a liquid, heat will flow by conduction according to the same formula as for a solid, but it will also flow by *convection.*

- In convection, material moves and carries heat from a hot region to a cold region.
- Typically, convection occurs because warmer liquid is less dense and rises, while cool liquid sinks. If a heat source is at the bottom of a pond, fluid will circulate, carrying heat to the surface with warm water, cooling at the surface, and returning cool water to the bottom.

Objects also exchange heat energy via electromagnetic waves. This mechanism is called *radiation.* The rate of energy transfer typically falls as the inverse of the square of the distance between the objects.

When the temperature of an object is increased, most objects expand. This effect is called thermal expansion.

- The length, L, increases by an amount ΔL according to the equation $\Delta L = L\alpha\Delta T$,where α is called the *thermal expansion coefficient*, and ΔT is the increase in the temperature of the object.

- The volume, V, increases by an amount ΔV according to the equation $\Delta V = V\beta\Delta T$ where $\beta \cong 3\alpha$ is called the *volume thermal expansion coefficient,* and ΔT is the increase in the temperature of the object.

Chapter 8

KINETIC THEORY AND THERMODYNAMICS

IDEAL GASES

An ideal gas is a dilute gas. Air is a reasonably good ideal gas. In atomic language, the gas is dilute enough that atoms of the gas rarely collide with each other.

Kinetic Model

In this model, a volume, V, of gas is composed of N atoms that do not interact with each other (except when rectifying a departure from equilibrium). The atoms collide elastically with the walls of the container, producing the observed pressure as the wall recoils from the collisions. Because there is no friction in the collision, there is no component of force parallel to the wall on an atom. Thus, the parallel (to the wall) component of atomic momentum is unchanged by the collision. The wall is far more massive than an atom so that in the elastic collision, the normal component of the atom velocity (the component perpendicular to the wall) is reversed in direction by the collision. (The magnitude of the velocity does not change.)

The average force by an atom on the wall is the impulse from one collision divided by the time it takes for the atom to return to the wall after bouncing off the opposite wall. The pressure, P, is calculated as

$$P = \frac{2}{3}\frac{N}{V}\left[\frac{1}{2}mv^2\right].$$

Ideal Gas Law

For an ideal gas, it is found experimentally that pressure, volume, and temperature are related by a simple *equation of state,* $PV = nRT$, where n is the number of moles of gas in the volume, R is a constant called the ideal gas constant, and T is the absolute temperature in degrees Kelvin. The conversion from Celsius to Kelvin temperature is

$$T(K) = T(°C) + 273K.$$

$T = 0K$ is a temperature called absolute zero.
The two equations for pressure agree with each other provided that

$$\frac{2}{3}N\left[\frac{1}{2}mv^2\right] = nRT.$$

$N\left[\frac{1}{2}mv^2\right]$ is the total kinetic energy of all the molecules in the gas. This is called the *internal energy* and given the symbol *U*.

It is verified experimentally that $U = \frac{3}{2}nRT$, so that kinetic theory, based on the atomic model, is experimentally vindicated.

THE FIRST LAW OF THERMODYNAMICS, INCLUDING PROCESSES ON PV DIAGRAMS

The internal energy can be increased by ΔU in two ways—the same two methods used in Chapter 7 to raise the temperature—add heat to the system, and do work on the system.

The result is $\Delta U = Q - W$, where the work, *W*, is the work done **by** the system. That is the reason for the sign in front of *W*. Work done by the system removes energy from it.

For an ideal gas contained in a cylinder with a movable piston, the work is $W = P\Delta V$, where ΔV is the increase in volume, provided that the pressure is constant. Then, the first law is $\Delta U = Q - P\Delta V$.

$P\Delta V$ is the area under a plot of pressure versus volume for the gas in the system. If the pressure is not constant, then $W = P\Delta V$ is incorrect, but *W* is still the area under that curve.

The figure shows some sample paths on the PV diagram for an ideal gas.

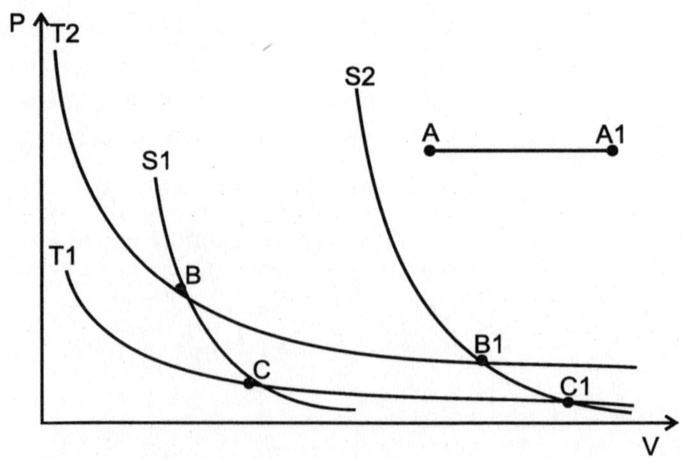

Every point on the graph represents a state of the gas. Each has a value of *P* and a value of *V*, and from the ideal gas equation of state, $PV = nRT$, we can find the temperature, *T*.

When two points on the graph (such as A and A1) are connected by a line, the line represents a *process* that carries the system from the first state to the second.

Because the work done by the system is the area under such a curve, we can use the graph to estimate work done. If the system is carried from A1 to A, the area is the same as the process taking it from A to A1. The difference is in the – signs.

When the system moves from A to A1, the volume expands and the system does work, so *W* is positive. When it moves from A1 to A, the volume contracts: Work is done upon the system. This means that *W* is *negative.*

A curve labeled T1 is shown. It represents a process that takes place without changing temperature, called an *isothermal process*. The curve labeled T2 represents another isothermal process, with T2>T1. If the temperature stays constant, then the internal energy also stays constant since for an ideal gas $U = \frac{3}{2}nRT$. If $\Delta T = 0$, the first law requires that $0 = \Delta U = Q - W$. That is, the heat energy in and the work out add to zero.

If the system moves from B to B1, all the heat input is converted to useful work. If it moves from C1 to C, all the work done on the system is converted to heat. On the upper path, heat must be added to the system to keep the temperature constant. On the lower path, heat must be removed to keep the temperature constant.

A curve labeled S1 is shown. This represents a process in which no heat flow is allowed, called an *adiabatic process.* The first law tells us that $-\Delta U = W$—work done by the system is equal to the decrease in internal—energy.

Along path B1 to C1, internal energy is completely converted to work. Along path C to B, work is converted to internal energy.

HEAT ENGINES

In the PV diagram on page 00, it is possible to select a sequence of paths that does two things:

- Returns the system to its original state, as if nothing had happened to it.
- Produces net work output. (i.e., the process converts heat to work). One such sequence is called the *Carnot cycle*, named because it is easy to analyze. It is the cycle B to B1 to C1 to C to B.

On the two adiabatic segments, the net work and the net change in internal energy is zero because the temperature difference is the same (but opposite) for the two segments. This means that the internal energy change is the same (but opposite) for the two segments and so is the work.

The (positive) work done on the hotter isothermal path is greater in magnitude than the (negative) work done on the colder isothermal path.

The net work for the cycle is the difference of the two isothermal works and is equal to the area inside the graph for the cycle.

SECOND LAW OF THERMODYNAMICS

The simplest statement of the second law of thermodynamics is this: When two isolated objects are in simple thermal contact, heat always flows from the hotter object to the colder one.

The law can be restated in terms of a parameter called the *entropy* of a system. In the same way that internal energy and heat were defined in terms of changes in those quantities, the entropy, S, is defined in terms of how it changes.

$\Delta S = \frac{\Delta Q}{T} = \frac{Q}{T}$ (Some books use ΔQ for heat flow and some use Q.)

In the example from page 00, the entropy of the hot object decreases while the entropy of the cold object increases. Because the cold temperature has smaller T, the entropy increase of the two-object isolated system shows a net increase.

This statement of the second law says that for an isolated system, the entropy always stays the same or increases. As a consequence, when the entropy is as large as possible, the system will stop changing.

As an example, consider a system composed of hot, moist air from above the Pacific Ocean and the cold Sierra Nevada Mountains. When the air reaches the mountains, heat flows from the air to the mountains. The cold water vapor condenses and forms ice crystals, the most highly ordered form of water. The entropy of the system increased. The entropy increase drove the process of ice formation.

It is sometimes said that when the entropy of a system increases, its *disorder* increases, so that the natural state of things is chaos.

In the atomic model for a gas, there are many different microscopic arrangements of atomic positions and velocities that produce the same macroscopic observations of P, V, and T. The microscopic state can change among many states without changing our observations.

The most probable macroscopic state is the one for which a change in microscopic state is least likely to cause an observable change in P, V, or T. Thus, the most stable macroscopic state is the one associated with the greatest number of microscopic states.

An equally valid statement of the second law is that when the system has a maximum number of microscopic *choices* the entropy is maximum and the macroscopic state is stable. In this way stability is associated with choice rather than disorder.

OTHER STATEMENTS OF THE SECOND LAW OF THERMODYNAMICS

The most efficient possible engine is the *Carnot* engine, which works with zero change in entropy.

The thermodynamic efficiency of an engine is the ratio of work output to energy input (as heat). It is a number between 0 and 1, with 1 representing complete conversion of heat to work.

$$\text{EFFICIENCY} = \eta = \frac{W_{CYCLE}}{Q_{INPUT}}$$

The efficiency of a Carnot engine is $\eta_{CARNOT} = \frac{W_{CYCLE}}{Q_{INPUT}} = 1 - \frac{T_{LOW}}{T_{HIGH}}$.

The temperature ratio is the ratio of the temperatures of the high and the low isotherms of the Carnot cycle.

- No process is possible whose sole outcome is the conversion of heat completely to work.

- No process is possible whose sole outcome is the transfer of heat from a cold object to a warmer object.

UNIT 3

Electricity and Magnetism

Chapter 9
ELECTROSTATICS

CHARGE, FIELD, AND POTENTIAL
CHARGE

The mathematical law that governs the force between two *charged* objects (We name the objects 1 and 2) has the same form as Newton's law for gravity.

In the case of gravity, the inverse-square law force is proportional to the product of the two masses— It depends on how much material is in each object.

The electrostatic force obeys a similar law, except that it depends on how much "electrical" material is in each object. This quantity is called the *electric charge* of an object. This quantity is given the symbol, q, and its unit is the *Coulomb*.

Charge in this sense is similar to the sense of the "charge" of gunpowder put into an old-fashioned cannon. The more charge—the stronger the force. Objects can become electrically charged without a noticeable change in their mass. (It turns out that adding charge to an object does increase its mass very slightly. In ordinary laboratory situations, electrical force associated with the added charge is billions of times larger than the change in gravitational force caused by the additional mass.)

Electrical charge can be moved from one object to another. If two objects are initially neutral and charge is moved from object 1 to object 2, then object 1 acts as if it were also charged. The charge on object 1 is opposite to the charge on object 2. That is, if each charge is brought near a third charge object (3), the electrostatic force on object 2 is equal but opposite to the force by object 3 on object 1.

There are only two kinds of charge, which are named "+" and "–" because these sign designations will become useful later. Opposite charges attract and like charges repel.

Charge is conserved. An object ordinarily has equal amounts of + and – electrical charge. If some – charge is removed, the object has an excess of + charge and is *positively charged.* If the – charge that was removed is placed on a second object, that object becomes *negatively charged.* Charge is not created or destroyed. It is moved about to create all the electrostatic effects that we see.

Charge adds. If two positive charges are placed on a piece of metal, the total charge on the metal is just the sum of the two charges. If one charge is negative and one positive, for addition use the – sign on the negative charge, so that the net charge on the metal is the difference between the two charges.

FIELD

An electrically charged object modifies the space around it by filling the space with an *electrical field.* If another charge encounters the electrical field, it feels a force determined by the field. There are many different arrangements of charges that can produce a desired field in a region of space. The force on a visiting charge depends only on the field and not on the details of the charges that created it.

. The electric field is a vector quantity, having magnitude and direction.

The magnitude of the force on a charge, q, is equal to the electric field magnitude, E, multiplied by q. If q is positive, the electric force on it is in the same direction as E. If q is negative, the electric force on it is opposite to the direction of E. Formally, $\vec{F}_{ELECTRIC} = q\vec{E}$ The unit of electric field is *Newton per Coulomb.*

POTENTIAL

The biggest virtue of the electric *potential* is that it can be easily measured in the laboratory. Charge and field can be measured, but there are no common charge meters or field meters. Potential is measured by a common instrument called a *voltmeter.*

It is commonly said that a voltmeter measures *voltage.* The proper term is *potential difference*. The unit of potential and potential difference is the *Volt.*

Like the electric field, the electric potential has a value at every point in space around a charge. The potential is a scalar instead of a vector. If a charge, q, is brought to the place where the potential has value V, then the electrical potential energy of that charge is $U_{ELECTRICAL} = qV$. Electric potential is electric potential energy per unit charge. One Volt is the same as 1 Joule per Coulomb.

If V is constant throughout a region of space, then that region has zero electric field. If V changes rapidly from one place to another, the electric field is strong. The potential difference determines the electric field strength. The electric field points in the direction of decreasing potential.

Calculus users, note that the electric potential difference between two points, A and B, is calculated from the electric field by

$$U_B - U_A = \Delta U_{AB} = -\int_A^B \vec{E} \bullet d\vec{s}$$

This integral is independent of path. The components of the electric field may be calculated from the potential using $E_x = -\frac{\partial U}{\partial x}$, etc. Because of this, the unit of electric field is often given as Volts per meter (equivalent to Newton per Coulomb).

INSULATORS AND CONDUCTORS

Some materials naturally hold charge in place. Objects that do not permit the charge that they possess to move about are called *insulators.* If a bit of charge is placed on an insulator, it stays in place. Real materials are not perfect insulators, but some, such as polystyrene and glass, are very good insulators.

Some materials allow charges that they possess to move freely. Objects that allow free-charge movement are called *conductors.* Metals are very good conductors. Some materials become extremely good conductors at low temperature. In this low temperature state they are called *superconductors.* Superconducting material have properties (such as the Meissner effect) that make them more complex than if they were simply perfect conductors.

Because of their conducting ability, the electric field inside a conductor (such as a metal) is zero. If charge is placed on a piece of metal, it arranges itself on the surface of the metal so that the electric field within is zero. All the charge placed on a metal resides on the surface. The inside of a metal is always neutral; the charge inside is zero. (This results from the metal having an equal number of electrons and protons so that the net charge is zero, even though there are many charges present.)

COULOMB'S LAW AND FIELD AND POTENTIAL OF POINT CHARGES

The magnitude of the force between two charges, q_1 and q_2, separated by a distance, *r*, is given by Coulomb's law: $F_{ELECTRIC} = k\frac{q_1 q_2}{r^2} = \frac{1}{4\pi\varepsilon_0}\frac{q_1 q_2}{r^2}$, where k is approximately $9\text{x}10^9 \frac{Newton\ m^2}{Coulomb^2}$ and

$\varepsilon_0 = 8.85\text{x}10^{-12}\frac{Coulomb^2}{Newton\ m^2}$.

The direction of the force is attractive if the charges are opposite. (If the charges are opposite, one is + and the other is – so that the product $q_1 q_2$ is negative.) The direction of the force is repulsive if the charges are the same. (If the charges are the same, then the product $q_1 q_2$ is positive.

The magnitude of the electric field due to a point charge *q*, as seen by an observer located a distance *r* from the charge, is $E = k\frac{q}{r^2} = \frac{1}{4\pi\varepsilon_0}\frac{q}{r^2}$.

For a positive point charge, the direction of the electric field is away from the point charge. For a negative point charge, the direction of the electric field is toward the point charge.

The electric potential of a point charge, q, is given by

$$V = k\frac{q}{r} = \frac{1}{4\pi\varepsilon_0}\frac{q}{r}.$$

- If the charge is positive, then the potential is positive.
- If the charge is negative, the potential is negative.
- Note that the potential goes to zero as the position of the observer goes to infinity.

The potential energy of a point charge, Q, in the presence of a point charge q is $U_{ELECTRIC} = k\frac{qQ}{r} = \frac{1}{4\pi\varepsilon_0}\frac{qQ}{r}$.

If the product qQ is negative, the potential energy is negative. If qQ is negative, the pair of charges can lower their energy by moving so that they become closer together.

If qQ is positive, the pair of charges lower their energy by moving farther apart. For calculus users:

- The radial component of the electric force on a charge Q in the presence of a charge q is (the negative of) the derivative of the potential energy with respect to r.
- The radial component of the electric field in the presence of a charge q is (the negative of) the derivative of the potential with respect to r.

FIELDS AND POTENTIALS OF OTHER CHARGE DISTRIBUTIONS

THE SINGLE MOST USEFUL CHARGE DISTRIBUTION IS THE *DIPOLE*.

A dipole consists of two charges of equal magnitude, q, and opposite sign, separated by a distance, d. The strength is measured by the dipole moment, $p = qd$. The direction of the dipole moment vector points from the negative charge towards the positive charge within the dipole charge pair.

The dipole electric field points away from the positive charge and towards the negative charge, "wrapping around" in space, from the positive to the negative.

The magnitude of the dipole electric field falls more rapidly with distance than the field from a single charge. The electric field strength is proportional to $\frac{1}{r^3}$, provided that r is much larger than the separation of the charges, d.

The electric potential at an observation point due to a dipole is given by $= \frac{1}{4\pi\varepsilon_0}\frac{p}{r^2}\cos\theta$, where r is the magnitude of the vector from the center of the dipole to the point of observation. θ is the angle between the vector $\vec{r}$ and the direction of the dipole moment vector, $\vec{p}$. As above, it is assumed that $r > d$.

PLANAR

ELECTRIC POTENTIAL OF A POSITIVE SHEET OF CHARGE

- A plane sheet of positive charge is a surface of constant potential.
- Above and below the sheet are other surfaces of constant potential.
- Each surface is a plane, parallel to the sheet of positive charge.
- The observed potential decreases as the point of observation is moved away from the sheet.
- Taking h as the distance from the sheet, the change in potential from one point to another is $\Delta V = V_2 - V_1 = -\frac{\sigma}{\varepsilon_0}[h_2 - h_1] = -\frac{\sigma}{\varepsilon_0}\Delta h$, where

    a. σ is the "surface charge density" of the sheet.
    b. it is equal to the total charge on the sheet divided by the area of the sheet.
    c. the units of σ are Coulombs per square meter.

ELECTRIC FIELD OF A POSITIVE SHEET OF CHARGE

- The electric field near a plane sheet of positive charge is perpendicular to the sheet. Above and below the sheet, the electric field points away from the sheet. The magnitude of the electric field is $\dfrac{\sigma}{2\varepsilon_0}$.

- The electric field is the same at all observation points near the sheet of charge. It is uniform.

- For a negative sheet of charge, the magnitudes are the same, but the potential increases as the observation point is moved away from the sheet and the electric field points towards the sheet (both above and below the sheet).

- Another planar charge distribution is the charge on the top surface of a large conducting sheet of metal, such as a sheet of aluminum foil. In order to make the electric field within the metal zero, the charge distributes equally on the top and the bottom of the metal sheet. The electric field is perpendicular to the sheet, pointing away from the sheet for a positively charged sheet. The magnitude of the electric field is $\dfrac{\sigma}{\varepsilon_0}$. Note the missing (compared to the field from a sheet of charge) factor of 2 in the denominator. This field is also uniform (independent of position) as long as the point of observation is near to the surface of the metal.

- A metal allows no variation in electrical potential throughout itself. A metal enforces a region of constant *V.*

SPHERICAL SYMMETRY

Outside of a spherical distribution of charge: If a distribution of charge has spherical symmetry, the electric field and the potential outside of the distribution is the same as if all the charge were concentrated at the center of the spherical distribution. The effective charge is the total charge in the distribution.

For an observation point within a spherically symmetric distribution of charge, the electric field and potential are the same as if they were caused by a charge at the center of the spherical distribution. The magnitude of the effective charge is equal to the charge contained within a sphere of radius *r* about the center of the distribution, where r is the distance of the observation from the center of the distribution. The electric field and the potential decrease to zero at the center of the distribution.

A charge surrounded by a spherical metal shell centered on the charge produces the ordinary point charge field, except within the metal itself. Inside the metal, the field is zero and the potential is constant. The potential outside the metal is the same as for a point charge. The potential just inside the shell is the same as the potential outside. This constant reduction in potential inside is applied to every potential inside.

For a charged solid metal sphere, all the charge resides on the spherical surface. The field and potential outside the sphere are the same as for a point charge at the center of the sphere. The effective charge is the net charge on the sphere. Inside the sphere the electric field is zero. The potential is constant and equal to the value just outside the surface of the sphere.

CYLINDRICAL SYMMETRY

For an observation point outside an infinitely long cylindrically symmetric distribution of charge, the electric field is radial (perpendicular to the axis of the cylinder).

- The electric field magnitude is given by $E = \dfrac{\lambda}{2\pi\varepsilon_0 r}$, where λ is the charge per unit length along the cylinder axis and r is the perpendicular distance from the observation point to the cylinder axis.
- For positive charge, the field points away from the cylinder axis. For an observation point outside an infinitely long cylindrically symmetric distribution of charge, the potential difference between two points is (–) the integral of the electric field:
- $\Delta V = V_2 - V_1 = -\dfrac{\lambda}{2\pi\varepsilon_0}\left[\ln(r_2) - \ln(r_1)\right] = \dfrac{\lambda}{2\pi\varepsilon_0}\ln\left[\dfrac{r_1}{r_2}\right].$
- For positive charge, the potential decreases as the distance to the cylinder axis increases.

GAUSS' LAW

For a small area element, $d\vec{A}$, the element of electric *flux*, $d\phi$, is defined as $d\phi = \vec{E} \bullet d\vec{A} = EdA\cos\theta$, where θ is the angle between $\vec{E}$ and the area vector, $d\vec{A}$. The direction of the area vector is perpendicular (normal) to the area itself. For a given electric field, the flux is greatest when the field is perpendicular to the area (i.e., $\vec{E}$ is parallel to $d\vec{A}$).

For a large surface, the flux through the surface is the integral

$$\phi = \int_{AREA} d\phi = \int_{AREA} \vec{E} \bullet d\vec{A} = \int_{AREA} EdA\cos\theta$$

Gauss law refers to a particular kind of flux: the flux through a *closed* surface. (A closed surface encloses a volume of space.) Gauss' law states that the flux through the closed surface is proportional to the total charge within the surface: $\int_{\substack{\text{AREA of} \\ \text{CLOSED} \\ \text{SURFACE}}} \vec{E} \bullet d\vec{A} = \frac{q_{INSIDE}}{\varepsilon_0}$, where the direction of $d\vec{A}$ is towards the *outside* of the enclosed volume.

For the special case of a point charge at the center of a sphere, the field is constant over the entire area and is radial (perpendicular to the area and parallel to $d\vec{A}$). This makes the integration easy. The result is identical to Coulomb's law for a point charge.

Gauss law is useful in cases such as above, in which $\vec{E} \bullet d\vec{A}$ is constant over the entire area. This can happen, for example, if

a. $\vec{E}$ is perpendicular to $d\vec{A}$ so that the integral is zero.
b. $\vec{E}$ is zero.
c. the surface is chosen to be a surface of constant electric field.
d. a combination of the above, for the various segments of the surface.

The key to using Gauss' law is that you get to choose the surface over which the integration is performed. If you choose the surface to match the symmetry of the charge distribution, you can make the integral easy to do. Typically the integral breaks into parts which are separately easy to do.

Chapter 10

CONDUCTORS, CAPACITORS, DIELECTRICS

ELECTROSTATICS WITH CONDUCTORS

The electrostatics of conductors is discussed in Chapter 9.

The most important fact to know is that a conducting object has the same potential at every point in the object. Charges and fields rearrange themselves to insure this result.

The interior of a conductor is neutral, and the interior electric field is zero. All excess charge resides on the surface of a conducting object.

Gauss' law may be used to show that, if a conducting object has a sharp point on it, the electric field outside the object, near to the point, is large. Charge rearranges itself on the surface to make this so.

CAPACITORS

PARALLEL PLATE

A parallel plate capacitor consists of two parallel metal plates, each of area *A*, separated by a distance *d*. Charge is removed from one plate and placed on the other. Each plate has a charge, *Q*. It may help to visualize the charge on the upper plate as $+Q$ and the charge on the lower plate as $-Q$.

The electric field from each plate has magnitude $\frac{\sigma}{2\varepsilon_0}$, where $\sigma = \frac{Q}{A}$ is the charge per unit area of one plate. The electric field due to the positive charge points away from the positive plate, and the electric field due to the negative charge points towards the negative plate. We neglect the "fringing" fields near the edges of the plate.

Since the fields from each plate are independent of distance from the plate (as long as the plate dimensions are large compared to the separation), the field between the plates is $\frac{\sigma}{\varepsilon_0}$, because the two fields add, pointing from the positive toward the negative plate. Outside the plates, the fields add to zero.

The magnitude of the potential difference between the plates is $\Delta V = Ed = \frac{Qd}{\varepsilon_0 A}$. This is usually rewritten as , $Q = \frac{\varepsilon_0 A}{d}\Delta V = C\Delta V = CV$, where *C* is called the capacitance of the pair of plates, and *V* refers to the potential difference, $V = \Delta V$. Such a pair of plates is called a *capacitor*.

The unit of capacitance is the Farad. $1\ Farad = 1\ F = \frac{1\ Coulomb}{Volt}$. In ordinary circuits, the capacitances used are typically measured in

$microfarad = \mu F = 10^{-6}\,F$.

The energy stored in a capacitor is $U = \frac{1}{2}QV = \frac{1}{2}CV^2 = \frac{1}{2}\frac{Q^2}{C}$

SPHERICAL

If two concentric metal spheres are used as the plates of a capacitor, we may visualize the outer sphere as the positively charged plate. The inner sphere has radius R_1 and the outer sphere has radius R_2.

The electric field outside the outer sphere is zero, by Gauss' law, since the two plates have equal but opposite charges.

The electric field between the two spheres is determined by the charge on the inner sphere. This field will point radially toward the center of the spheres, and Gauss' law tells us that its magnitude will be $E = \frac{Q}{4\pi\varepsilon_0 r^2}$ where *r* is the distance of the observation point from the center of the two spheres.

The potential difference between the two spheres is the (negative) integral of the electric field: $\Delta V = -\int_{R_1}^{R_2} \vec{E} \bullet d\vec{r} = \frac{Q}{4\pi\varepsilon_0}\left[\frac{1}{R_1} - \frac{1}{R_2}\right]$. The potential of the positive outer plate is larger than the potential of the negative inner plate.

The capacitance of this spherical capacitor is

$$C = \frac{Q}{\Delta V} = 4\pi\varepsilon_0\left[\frac{1}{R_1} - \frac{1}{R_2}\right]^{-1}.$$

Two special cases are of interest:

1. If the outer sphere is made very large, the second term vanishes. A single sphere has a capacitance $C_{\substack{SINGLE\\SPHERE}} = 4\pi\varepsilon_0 R_1$.
2. If the separation of the plates, *d*, is small compared to the radii of the spheres, then $C_{CLOSE} = \frac{4\pi\varepsilon_0 R^2}{d}$, where *R* is the average radius of the two spheres.

CYLINDRICAL

If two concentric metal cylinders are used as the plates of a capacitor, we may visualize the outer cylinder as the positively charged plate. The inner cylinder has radius R_1 and the outer cylinder has radius R_2. Both cylinders are very long compared to their radii.

The electric field outside the outer cylinder is zero, by Gauss' law, since the two plates have equal but opposite charges.

The electric field between the two cylinders is determined by the charge on the inner cylinder. This field will point radially toward the central axis of the cylinders, and Gauss' law tells us that its magnitude will be

$E = \dfrac{Q/L}{2\pi\varepsilon_0 r}$, where L is the length of the cylinders and r is the distance of the observation point from the central axis of the two cylinders.

The potential difference between the two cylinders is the negative integral of the electric field: $\Delta V = -\int_{R_1}^{R_2} \vec{E} \bullet d\vec{r} = \dfrac{Q/L}{2\pi\varepsilon_0} \ln\left[\dfrac{R_2}{R_1}\right]$. The potential of the positive outer plate is larger than the potential of the negative inner plate.

The capacitance of this cylindrical capacitor is $C = \dfrac{Q}{\Delta V} = \dfrac{2\pi\varepsilon_0 L}{\ln\left[\dfrac{R_2}{R_1}\right]}$.

DIELECTRICS

Insulating materials are materials that do not allow free motion of charges placed on them. Insulators do, however, allow motion of charges within the molecules from which they are made.

An electric field applied to the material pulls + charges one way and – charges the other. As a result, dipole moments are created whose dipole moment is proportional to the strength of the applied electric field.

If the material is placed between the plates of a capacitor, the net effect is a shift of negative charge to the surface of the material nearest the + charged plate and positive charge to the surface nearest the – charged plate. This charge distribution produces an electric field within the material that tends to cancel the applied electric field.

Because of this "anti-electric field" response, the material is called a *dielectric* material.

Since, for a given charge, Q, on the plates, the dielectric material reduces the electric field, the dielectric material also reduces the potential difference between the two plates. The capacitance of the capacitor is, therefore, increased. The capacitance, C, with the dielectric present, is related to the empty capacitance, C_0, by $C = \frac{Q}{\Delta V} = \varepsilon C_0$, where ε is called the dielectric constant of the material.

The *dielectric constant* is always greater than 1 and is typically less than 10, although the dielectric constant for water is about 80. This large value reflects the highly polar nature of the water molecule.

Chapter 11
ELECTRIC CIRCUITS

CURRENT, RESISTANCE, AND POWER

CURRENT

The flow of charge from one place to another is called a *current*. If an observer at a point in space sees charge Δq move past that point in time Δt, the current, *I*, is measured by the rate of that motion: $I = \frac{\Delta q}{\Delta t} \rightarrow \frac{dq}{dt}$ is called the current.

The unit of current is the Ampere, abbreviated Amp or A.

$Amp = \frac{Coulomb}{second}$. Observed currents can ordinarily range from hundreds of Amps to milliAmps (mA) to microAmps (μA).

RESISTANCE

In ordinary conductors, some energy from the moving charges in a current is converted to heat, so that energy is extracted from the flow. This characteristic of materials is called *electrical resistance.*

In practice, the current is the same all along the conductor, and the heat energy transfer is reflected in a decrease in electrical potential energy. This, in turn, is reflected in a decrease in electrical potential, *V*, along the direction of current flow.

In many practical situations, the size of the current flow through an object is proportional to the decrease in V across the object. The constant of proportionality is called the resistance, *R*: $\Delta V_{ACROSS} = I_{THROUGH} R$ or, more compactly, $V = IR$. This relation is called *Ohm's law*. The unit of resistance is the Ohm, symbolized by Ω. Clearly, $\Omega = \frac{Volt}{Ampere}$. Resistances in ordinary objects range from microOhms ($\mu\Omega$) to MegOhms (MegaOhms) ($M\Omega$). The common unit in electrical circuits is the kiloOhm, ($k\Omega$).

POWER:

The rate at which electrical potential energy is converted to heat is the power, P, dissipated in the object. It is given by $P = I_{THROUGH} \Delta V_{ACROSS}$ or more simply as $P = IV$. Ohm's law can be used to show that $P = IV = I^2R = \frac{V^2}{R}$. The unit of power is the Watt: $Watt = \frac{Joule}{Second}$.

STEADY-STATE DIRECT CURRENT CIRCUITS WITH BATTERIES AND RESISTORS ONLY

Most electrical devices require a flow of current and a transfer of energy derived from electrical potential energy. The simplest systems with these two characteristics combine batteries (as a source of electrical potential energy) with objects that obey Ohm's law (called resistors).

A battery may be regarded as a charge pump, moving positive charges from one terminal (the – terminal) to the other (the + terminal) and increasing the electrical potential energy of each charge that it moves. A battery neither creates nor destroys charge.

- The electrical potential energy per Coulomb of charge moved is the potential difference observed across the terminals of the battery.
- The potential difference across the terminals is loosely called the Voltage of the battery.
- If no current is flowing in the battery, the potential difference across the terminals is called the EMF of the battery. The EMF is given the symbol, $\mathcal{E}$, to separate it from "passive" potential differences. An EMF can supply power, while passive objects only receive (and transmit) power.
- When current flows, the battery acts as if it has an internal resistance so that the observed battery voltage is less than the EMF. The difference is the product of the current with the internal resistance.

If a resistor is connected across the battery terminals, + charges from the + (high potential energy) terminal flow through the resistor to the low potential energy (–) terminal, losing their potential energy to heat, which remains in the resistor.

More complicated circuits require schematic diagrams, using symbols to present a compact description of the circuit. The symbols and terms commonly used are defined in the following figures.

The figure below shows the simplest circuit, with a battery "pushing" a current through a resistor. An ammeter is inserted to measure the current flowing through both the resistor and the battery. A voltmeter is connected above and below the resistor to measure the potential difference across the resistor.

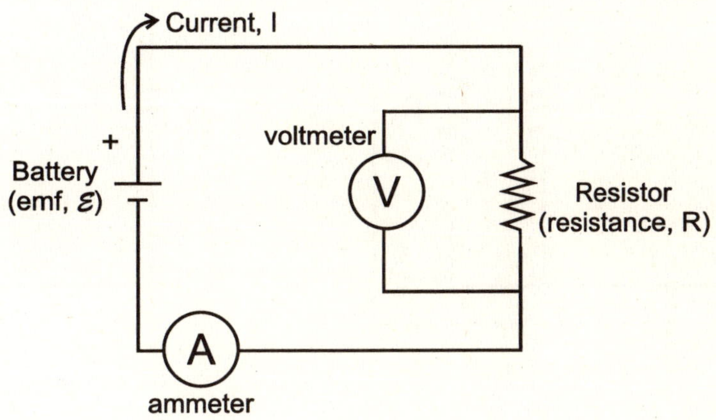

The figures below show two circuits with two resistors. In one circuit the resistors are in parallel, and in the other the resistors are in series. The current through the battery for each combination of resistors can be calculated as though the combination had been replaced by a single "equivalent" resistor.

Series Resistors

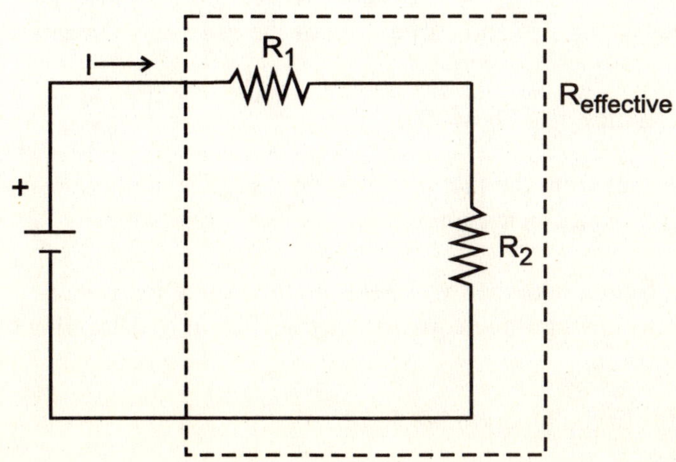

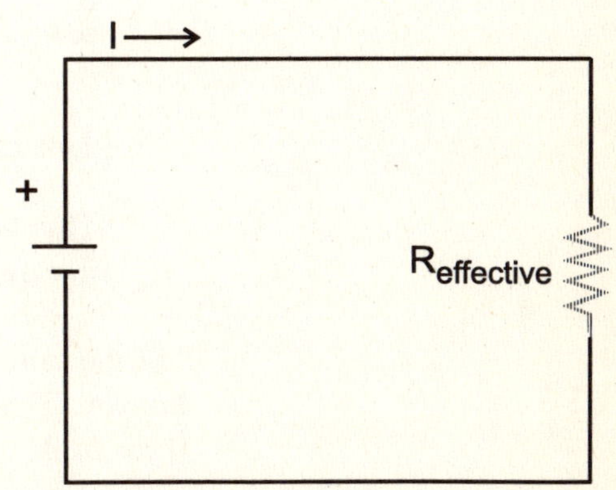

$R_{effective} = R_1 + R_2$

Parallel Resistors

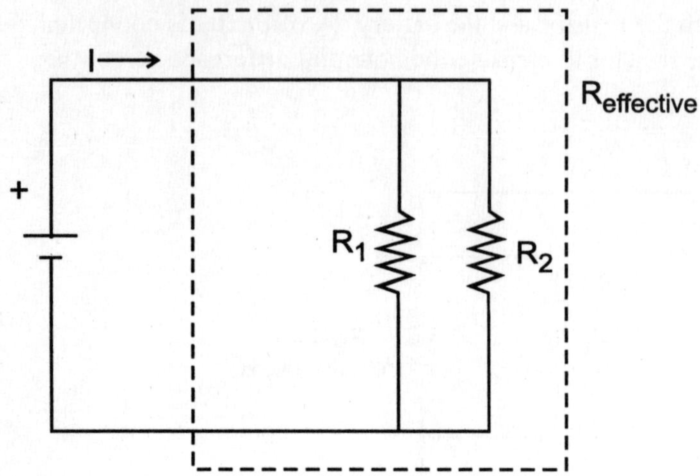

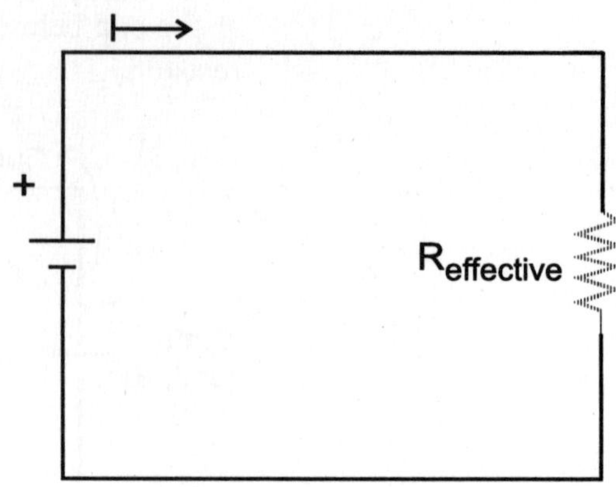

$$\frac{1}{R_{effective}} = \frac{1}{R_1} + \frac{1}{R_2}$$

Kirchhoff's laws provide a mechanism to analyze any circuit. The two Kirchhoff's laws are:

1. $\{\Delta V_1 + \Delta V_2 + \Delta V_3 + \ldots\}_{LOOP} = 0$. The sum of the potential differences around a loop must be zero. Some potential differences must be positive and some negative. This law derives from the conservative nature of the electric force and the fact that the electric potential energy at a particular point in the circuit is unique, independent of the path.
2. $\{I_1 + I_2 + I_3 + \ldots\}_{JUNCTION} = 0$. where current into the junction of several wires is positive and current out of the junction is negative.

   This equation may also be written $\left\{I_{\substack{INTO\\JUNCTION}}\right\}_{NET} = \left\{I_{\substack{OUT\ OF\\JUNCTION}}\right\}_{NET}$.
   This law derives from the fact that electric charge is neither created nor destroyed in an electric circuit.

For the following figure, take the two EMFs and the three resistances as known. Then the following three equations are sufficient to determine the three (unknown) currents.

$I_1 = I_2 + I_3$ for the junction labeled J.

$\mathcal{E}_1 - I_1R_1 - I_2R_2 = 0$ for the loop labeled A.

$I_2R_2 + \mathcal{E}_2 - I_3R_3 = 0$ for the loop labeled B.

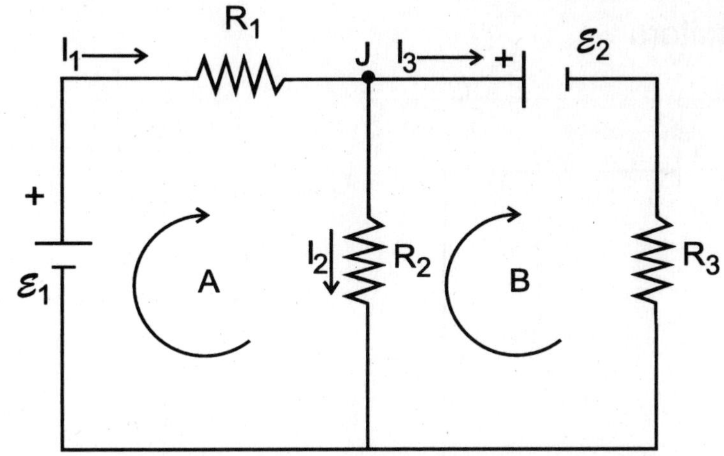

Kirchhoff's laws can be used to derive the formulas given above for series and parallel resistors.

CAPACITORS IN CIRCUITS

STEADY STATE

The figure below shows a simple capacitor circuit. The charge on the capacitor, Q, is related to the voltmeter reading, V, by $Q = CV$

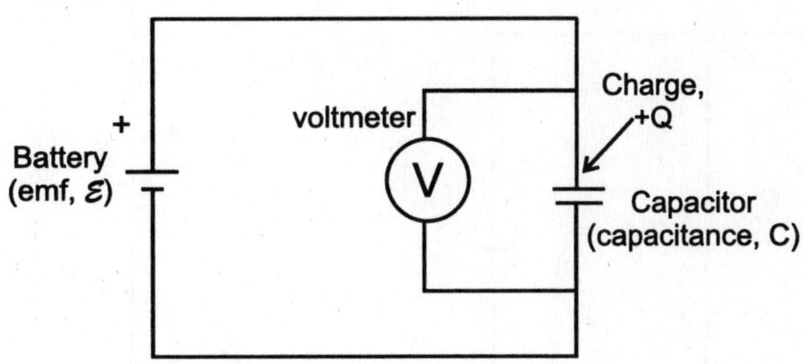

The figures below show series and parallel combinations of capacitors. Kirchhoff's loop law for the two circuits gives $\mathcal{E} - \frac{Q}{C_1} - \frac{Q}{C_2} = 0$ for the series capacitors and $\mathcal{E} = \frac{Q_1}{C_1} = \frac{Q_2}{C_2}$ and $\frac{Q_1}{C_1} - \frac{Q_2}{C_2} = 0$ for the parallel capacitors.

Series Capacitors

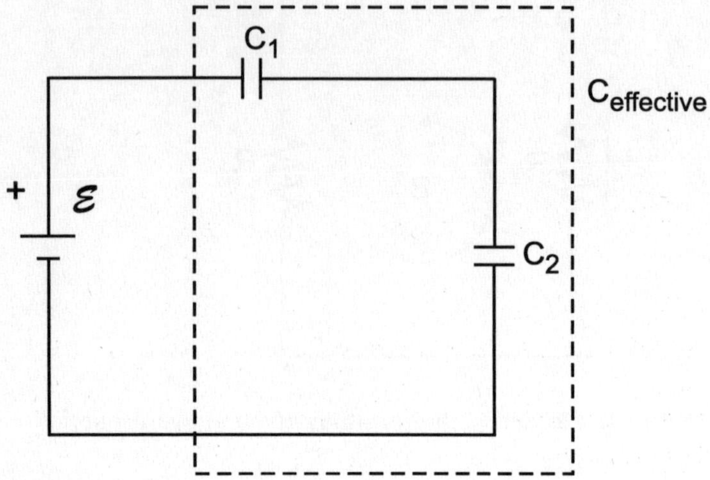

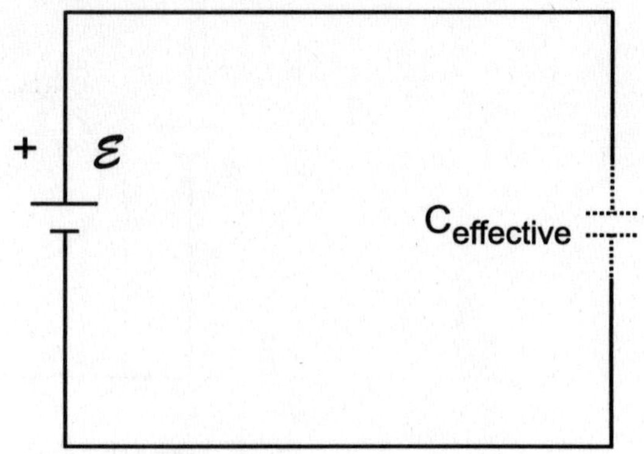

$$\frac{1}{C_{effective}} = \frac{1}{C_1} + \frac{1}{C_2}$$

Parallel Capacitors

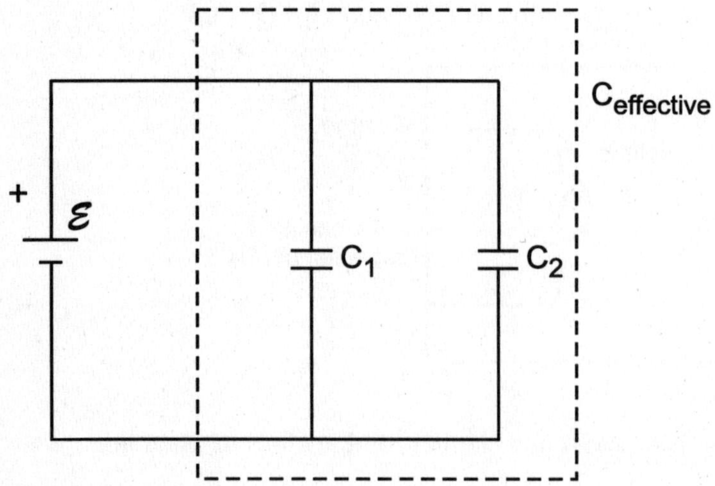

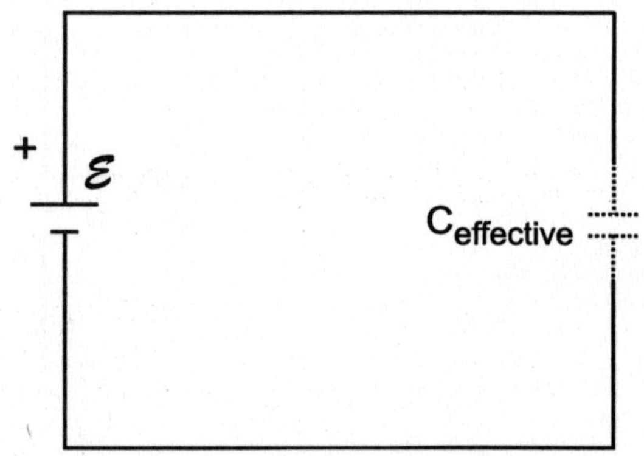

$$C_{effective} = C_1 + C_2$$

Transients in RC Circuits

In the circuit below, the switch represented by the arrow is closed at time t=0. Kirchhoff's loop law gives $\frac{Q}{C} - iR = 0$, for t>0. The lower case i is used to represent current, which can change as time goes by.

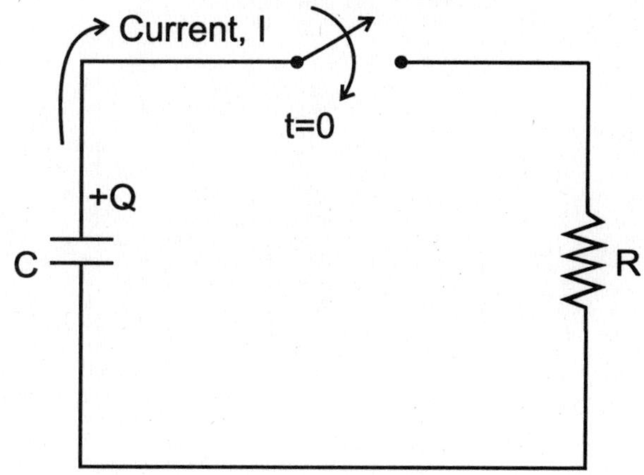

The current is related to the charge on the capacitor by $i = -\frac{dQ}{dt}$. The solution to the resulting differential equation for the charge on the capacitor is $Q = Q_0 e^{-t/\tau}$, where Q_0 is the initial charge on the capacitor and $\tau = RC$ is the characteristic decay time. Current and capacitor voltage obey similar exponential decay equations.

In the circuit below, the switch represented by the arrow is closed at time t=0. Kirchhoff's loop law gives $\mathcal{E} - \frac{Q}{C} - iR = 0$, for t>0.

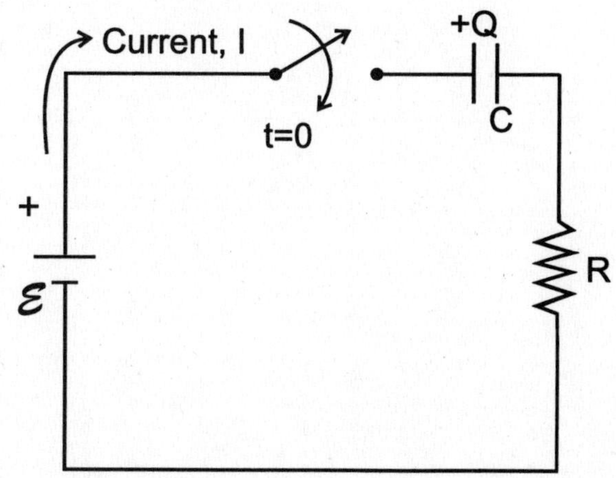

The current is related to the charge on the capacitor by $i = +\frac{dQ}{dt}$. The solution to the resulting differential equation for the charge on the capacitor is $Q = Q_\infty \left[1 - e^{-t/\tau}\right]$, where $Q_\infty = C\mathcal{E}$ is the final charge on the capacitor and $\tau = RC$ is the characteristic time. Capacitor voltage obeys a similar exponential equation.

Chapter 12
MAGNETOSTATICS

A compass may be used to probe magnetic fields. The compass needle points in the direction of the magnetic field vector, $\vec{B}$. The strength of the torque that aligns the compass needle is proportional to the magnitude of the magnetic field, $B = |\vec{B}|$. The strength of the magnetic field is measured in units of *Tesla,* abbreviated *T.*

FORCES ON MOVING CHARGES IN MAGNETIC FIELDS

The force on a charge, *q,* moving at velocity, $\vec{v}$, in a uniform magnetic field, $\vec{B}$, is given by $\vec{F} = q\vec{v} \times \vec{B}$. The vector product describes the fact that the force is perpendicular to both the velocity and the magnetic field vectors.

For positive charges, the direction of the force is given by the usual vector product right hand rule. For a negative charge, the force is opposite to that for a positive charge.

A charge of mass, *m,* moving perpendicular to a uniform magnetic field (with no other forces present) will accelerate in a direction perpendicular to both the velocity and the field. It will execute circular motion. The radius of the circle is given by $R = \frac{mv}{qB}$. *R* is called the radius of the cyclotron orbit.

If a component of the velocity is parallel to the magnetic field, that component will experience no change. The velocity component parallel to $\vec{B}$ causes the path of the charge to be a helix, with circular motion in the plane perpendicular to the field.

TORQUE AND FORCE ON MAGNETIC DIPOLES IN MAGNETIC FIELDS

The magnetic *dipole* is not made up of a pair of magnetic charges. Experimentally, the *dipole* is the most fundamental magnetic entity. No magnetic charge (called the *magnetic monopole*) has ever been experimentally identified. The dipole is given its inconsistent name because its magnetic field has the same fountain-like shape as the electric dipole field. The dipole moment is given the symbol $\rightarrow$.

In a uniform magnetic field, a dipole experiences no force. It does experience a torque, which tends to align the dipole moment with the external field. The torque is given by $\vec{\tau} = \vec{p} \times \vec{B}$.

In a nonuniform magnetic field, the dipole still tends to align parallel to the field. It also is pulled towards the region of stronger magnetic field. It is this force by a nonuniform magnetic field that makes magnets pull towards each other.

FORCES ON CURRENT-CARRYING WIRES IN MAGNETIC FIELDS

Consider an electric current traveling in a segment of wire of length *D*, composed of *N* positive charges, *q*, moving at average velocity, *v*. We imagine that the charge moves from left to right. The moving charge in this segment, $\Delta Q = Nq$, needs a time $\Delta t = \frac{D}{v}$ to completely pass out of the right hand end of the segment. An observer at that point calculates a current of $I = \frac{\Delta Q}{\Delta t} = \frac{Nqv}{D}$.

If the segment of wire is in a magnetic field, $\vec{B}$, the force on the wire is the same as the force on the moving charges in the segment, $\vec{F} = Nq\vec{v} \times \vec{B}$. The magnitude of the force on a straight wire segment of length *D* is $F = NqvB\sin\theta = IBD\sin\theta$, where θ is the angle between the direction of current flow and the direction of the magnetic field. The direction of the force is given by the right hand rule.

FIELDS OF LONG CURRENT-CARRYING WIRES

The observed magnetic field due to a long straight wire carrying a current, *I*, has magnitude $B = \frac{\mu_0 I}{2\pi r}$, where r is the distance from the wire to the observation point (measured perpendicular to the wire). μ_0 is a constant called *the permeability of free space*. $\mu_0 = 4\pi \text{x} 10^{-7} \frac{Tesla}{meter\ Ampere}$.

The direction of the magnetic field is perpendicular to the radius vector that runs perpendicular from the wire to the point of observation. There is no component of magnetic field parallel to the wire. The magnetic field vectors lie on circles around the wire with their centers at the center of the wire. If the right hand grasps the wire with the thumb in the direction of the current, the fingers curl around the wire in the same direction as the magnetic field.

Two long parallel wires (carrying currents I_1 and I_2) exert a force on each other, mediated by this magnetic field. If the currents flow in the same direction, the force between segments of length D is attractive and has magnitude $F = \dfrac{\mu_0 I_1 I_2 D}{2\pi r}$, where r is the separation of the wires. If the currents flow in opposite directions the force is repulsive, with the same magnitude. The magnetic force tends to pack currents together.

BIOT-SAVART LAW AND AMPERE'S LAW

Biot-Savart Law

The *Biot-Savart* law allows calculation of the magnetic field for a wire of any shape.

First, we calculate the magnetic field due to an infinitesimal length of wire, ds: $d\vec{B} = \dfrac{\mu_0}{4\pi}\dfrac{Id\vec{s}\times\vec{r}}{r^3}$. Note that this field falls proportional to the distance squared from the bit of wire. The direction of $d\vec{s}$ is the same as the direction of the current in the wire.

Second, at a particular observation point (called a "field point"), the contribution from each bit of wire, $d\vec{s}$, is computed as

$$\vec{B} = \frac{\mu_0}{4\pi}\int_{\substack{ALL\\CURRENT}} \frac{Id\vec{s}\times\vec{r}}{r^3}.$$

This method is often useful in practical calculations of magnetic field.

Ampere's Law

Ampere's law is most useful in highly symmetrical situations.

Ampere's law also involves an integral over a vector element, $d\vec{s}$, but now the element is along a closed loop in the region where the field is to be calculated, rather than being associated with the currents that cause the field. For a closed loop, the integral of the magnetic field along the loop obeys this equation: $\oint_{LOOP}\vec{B}\bullet d\vec{s} = \mu_0 I_{\substack{THROUGH\\THE\ LOOP}}$. In order to understand the current "through the loop" picture a piece of clear plastic-wrap that is laid across the loop, "sealing" it as if the loop marked the top of a jar. Current through the loop is the *net* current, which penetrates the plastic wrap.

Current through the loop can be positive or negative. If the fingers of the right hand are placed along the loop, pointing in the same direction as $d\vec{s}$, then the right thumb points in the direction of positive current through the loop.

All of the currents through the loop are summed, including their + and – signs to find $I_{\substack{THROUGH \\ THE\ LOOP}}$.

Ampere's law allows easy calculation of the magnetic field due to a long straight wire, for which the appropriate loop is a circle with its center on the axis of the wire. It does not work easily for the field due to a single loop of conducting wire.

The relation between Ampere's law and the Biot-Savart law is similar to the relation between Gauss' law and Coulomb's law for electrostatics. It turns out that Ampere's law and Gauss' law are the more fundamental.

Chapter 13

ELECTROMAGNETISM

ELECTROMAGNETIC INDUCTION (INCLUDING FARADAY'S LAW AND LENZ'S LAW)

Faraday's law is stated in two steps. Refer to the loop and magnetic field shown below. The magnetic flux through a loop of area, A, is defined as the product $\phi = AB\cos\theta$.

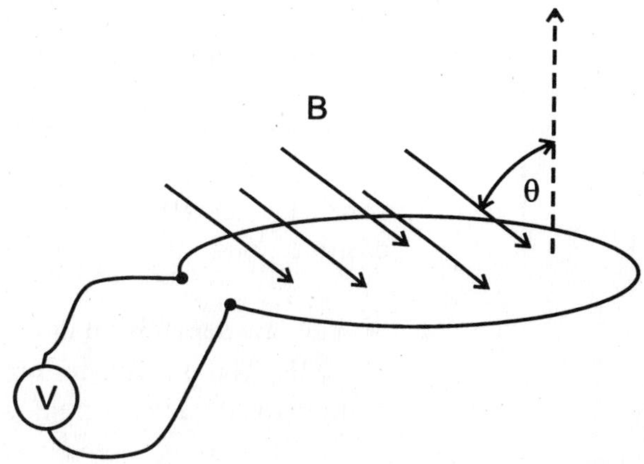

The angle, θ, is the angle between the magnetic field and the direction perpendicular (normal) to the loop area. (Calculus users, note that the formal definition of flux is $\phi = \int_{AREA\ OF\ LOOP} \vec{B} \bullet d\vec{A}$.)

The experiment shows that, for a loop of wire as shown, a potential difference is produced when the magnetic flux through the loop changes. If the flux stops changing, the potential drops to zero. Since this potential is not passive (it is capable of delivering power), it is called an EMF, following the practice for a battery.

The EMF is said to be *induced* by a changing flux and is simply $\mathcal{E}$ = (rate of change of)ϕ.

- If a plot of flux versus time makes a straight line, the rate of change of flux (and the induced EMF) is simply the slope of the straight line.

- Calculus users, note that the EMF is given by $\varepsilon = (-)\frac{d\phi}{dt}$. The minus sign is associated with Lenz's law; see below.

- Calculus users note that the potential difference between the location of the two terminals of the loop of wire can be calculated even if the wire is not present. The potential difference is the integral along a path following the loop path: $(-)\frac{d}{dt}\int_{\substack{AREA\ OF\\ LOOP}} \vec{B} \bullet d\vec{A} = (-)\frac{d\phi}{dt} = \mathcal{E} = \oint_{LOOP} \vec{E} \bullet d\vec{s}$. The minus sign now makes sense: Placing the fingers of the right hand along the direction of $d\vec{s}$, the right thumb points in the direction of $d\vec{A}$.

Lenz's law shows how to determine which terminal of the loop will be positive.

- The law depends on the current that flows in the wire to produce the EMF. The positive terminal will be made positive by current flowing towards it. This current is called the *induced current.*

- The induced current creates, as if flows, its own magnetic field near the loop. This field adds with the original magnetic field, but it is useful to think of it separately.

- Lenz's law states that the induced magnetic field acts to *oppose the change* in flux of the original field.

The flux that induces the EMF can change in several interesting ways:

- The original magnetic field changes magnitude. This is the basis of the electrical transformer.

- The original magnetic field changes direction. This is used in some magnetic navigation sensors.

- The loop changes direction. Also used in magnetic direction sensors, and in magnetic field strength sensors.

- The loop changes size. If the sides of the loop are square, we can imagine the loop growing so that one straight side moves perpendicular to its length. The work done by the EMF created this way is the same as would be calculated using the force calculated from $\vec{F} = q\vec{v} \times \vec{B}$.

Together, Faraday's law and Lenz's law take the place of the simple force equation above, for the magnetic force on a moving charge.

INDUCTANCE (INCLUDING LR AND LC CIRCUITS)

INDUCTANCE

A coil of wire with *N* turns (loops) of cross sectional area, *A,* experiences a flux $\phi = NAB$ when a uniform field, *B*, is applied parallel to the axis of the coil.

If the magnetic field is changing with time, an EMF is induced in each loop. The net EMF from one end of the coil to the other is the sum of the individual EMF's. $|\mathcal{E}| = \left|\frac{d\phi}{dt}\right|_{COIL} = NA\left|\frac{dB}{dt}\right|$.

It is possible that the magnetic field above is caused by the same coil that is producing the induced EMF. To make this happen, a changing current must be passed through the coil. The magnetic field is proportional to the current flowing through the coil. The rate of change of flux is proportional to the rate of change of current.

In this case, the induced EMF is proportional to the rate of change of the current, *i,* in the coil. The constant of proportionality is called the inductance, *L*: $|\mathcal{E}| = L\left|\frac{di}{dt}\right|$. The unit of inductance is the Henry,

$$Henry = \frac{Volt\ second}{Ampere}.$$

The energy stored in an inductor is $U_L = \frac{1}{2}Li^2$.

LR CIRCUITS

When a coil is used in a circuit whose current can vary, the potential difference across the coil is a passive response to the changing current. Because the coil is passive and not a source of energy, the potential difference is represented by ΔV or V, just as for a resistor or a capacitor. Lenz's law is used to determine the sense of the voltage. The potential across the coil acts in a sense to oppose the change in current: If the current is increasing, the coil potential acts to prevent the increase.

In the circuit, the experiment begins when the switch is thrown so that current must flow in the resistor-inductor circuit, without going through the battery.

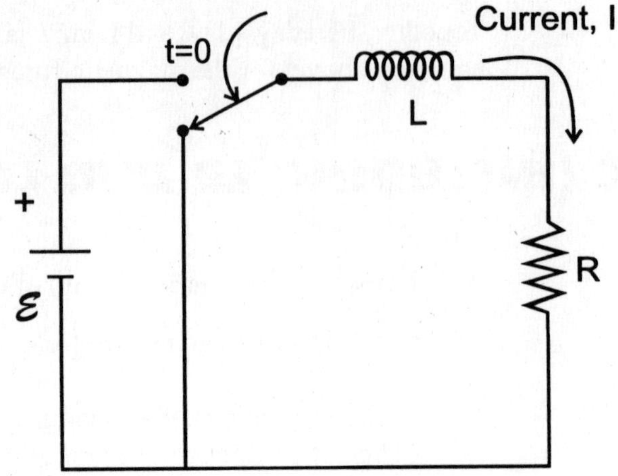

An EMF is induced in the coil that opposes the tendency of the current to fall. Kirchhoff's loop law for this circuit is $-L\frac{di}{dt} - iR = 0$. The solution to this equation is $i = i_0 e^{-t/\tau}$, where $i_0 = \frac{\varepsilon}{R}$ is the initial current and the decay time is $\tau = \frac{L}{R}$. If at t=0 the switch is thrown the other way, connecting to the battery, the current is given by $i = i_\infty\left(1 - e^{-t/\tau}\right)$ where $i_\infty = \frac{\varepsilon}{R}$ is the final steady state current.

LC Circuits

In the figure below, a capacitor starts with an initial charge, Q_0, and the switch is closed, allowing current to flow. Kirchhoff's law for this loop is

$$L\frac{di}{dt} - \frac{Q}{C} = 0.$$

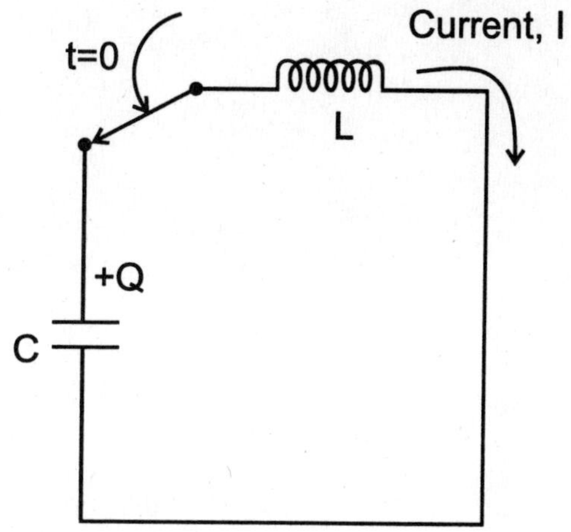

The relation between current and charge is $i = -\frac{dQ}{dt}$, making the Kirchhoff equation $\frac{d^2Q}{dt^2} = -\frac{Q}{LC}$. This has the form of the equation for simple harmonic motion. The solution for this problem is $Q = Q_0 \cos \omega_0 t$, where Q_0 is the initial charge on the capacitor and $\omega_0 = \frac{1}{\sqrt{LC}}$ is the natural frequency of the LC oscillator.

The current for this solution is $i = -\frac{dQ}{dt} = i_{MAX} \sin \omega_0 t$ where $i_{MAX} = \omega_0 Q_0$.

The energy stored in this oscillator is $U_L = \frac{1}{2}L(i_{MAX})^2 = \frac{1}{2}\frac{Q_0^2}{C}$. The energy oscillates between storage in the electric field of the capacitor and the magnetic field of the inductor.

MAXWELL'S EQUATIONS

All of the electrical and magnetic experimental results reported above can be summarized in four equations:

1. $$\int_{\substack{\text{AREA of}\\\text{CLOSED}\\\text{SURFACE}}} \vec{E} \bullet d\vec{A} = \frac{q_{INSIDE}}{\varepsilon_0}$$ (Gauss' law)

2. $$\oint_{LOOP} \vec{E} \bullet d\vec{s} = (-)\frac{d}{dt} \int_{\substack{\text{AREA OF}\\\text{LOOP}}} \vec{B} \bullet d\vec{A}$$ (Faraday's law)

3. $$\int_{\substack{\text{AREA of}\\\text{CLOSED}\\\text{SURFACE}}} \vec{B} \bullet d\vec{A} = 0$$ (Gauss' law for magnetism, reflecting the fact that no magnetic monopoles are observed.)

4. $$\oint_{LOOP} \vec{B} \bullet d\vec{s} = \mu_0 I$$ (Ampere's law)

Maxwell added a term to Ampere's law to make the equations cover more situations, giving a set of equations that are called Maxwell's equations.

1. $$\int_{\substack{\text{AREA of}\\\text{CLOSED}\\\text{SURFACE}}} \vec{E} \bullet d\vec{A} = \frac{q_{INSIDE}}{\varepsilon_0}$$ (Gauss' law)

2. $$\oint_{LOOP} \vec{E} \bullet d\vec{s} = (-)\frac{d}{dt} \int_{\substack{\text{AREA OF}\\\text{LOOP}}} \vec{B} \bullet d\vec{A}$$ (Faraday's law)

3. $$\int_{\substack{\text{AREA of}\\\text{CLOSED}\\\text{SURFACE}}} \vec{B} \bullet d\vec{A} = 0$$ (Gauss' law for magnetism, reflecting the fact that no magnetic monopoles are observed.)

4. $$\oint_{LOOP} \vec{B} \bullet d\vec{s} = \mu_0 I + \mu_0 \varepsilon_0 \int_{\substack{\text{AREA OF}\\\text{LOOP}}} \vec{E} \bullet d\vec{A}$$ (modified Ampere's law)

These equations may be solved in a region of space where no electric charges and no magnetic dipoles are found. The most interesting solution is a wave solution, called an *electromagnetic wave*.

The wave travels at a speed, $c = \sqrt{\frac{1}{\mu_0 \varepsilon_0}}$. With proper choice of *xyz* coordinate system, the electric component of the wave is all along the *y*-axis and is written $E_y = E_0 \cos(\omega t - kx)$, where $\omega = 2\pi f = \frac{2\pi}{T}$ is the angular frequency, f is the ordinary frequency, and T is the period of oscillation of the electric field, as measured at a point of observation, as the wave passes by. $k = \frac{2\pi}{\lambda}$ is the wave vector or wave number, and λ is the wavelength, the spatial period of the wave.

The magnetic field is all along the z-axis and is "in phase" with the electric field: $B_z = B_0 \cos(\omega t - kx)$

The electric and magnetic field amplitudes are related by $\frac{E_0}{B_0} = c$

The wave carries both momentum and energy. The energy per unit volume is given by $\begin{pmatrix} Average \\ Energy \\ Density \end{pmatrix} = \frac{1}{2}\varepsilon_0 E_0^2$. The momentum per unit volume is given by $\begin{pmatrix} Average \\ Momentum \\ Density \end{pmatrix} = \frac{\frac{1}{2}\varepsilon_0 E_0^2}{c}$.

UNIT 4
Waves and Optics

Chapter 14
WAVE MOTION

PROPERTIES OF TRAVELING WAVES

Any disturbance that maintains its shape and travels at a velocity, *v,* is a *traveling wave*. The waves most commonly described are *sinusoidal*, having the form shown in the figure below.

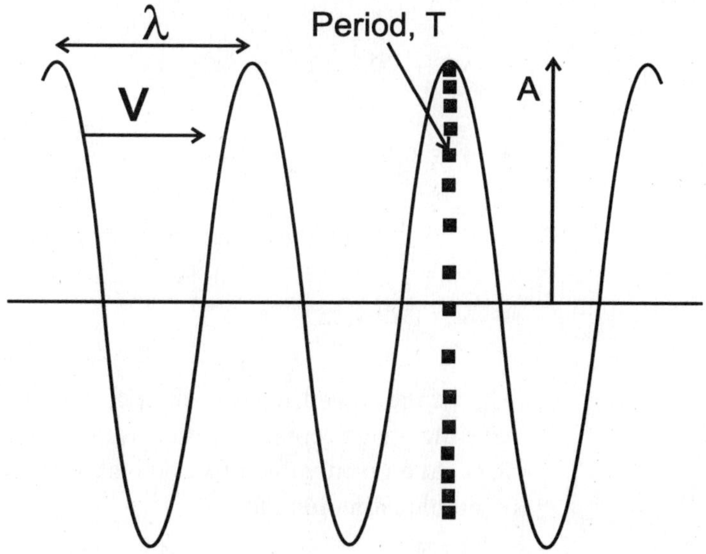

The horizontal line in the figure represents the equilibrium position of the medium. It may be the undisturbed surface of a lake. The wave displaces up and down from equilibrium by equal amounts. The maximum vertical displacement is the *amplitude, A.* The wave travels from left to right with velocity, *v.*

A single point in the medium, represented by a square dot, moves up and down as the wave passes. Each time a crest passes, the point reaches its highest displacement. The motion of the point is periodic. It repeats after a time, *T*, and is called the *period* of the wave.

The separation of the wave crests is called the *wavelength*, symbolized by the Greek letter, λ.

During one period, T, a wave crest advances a distance equal to λ.

The speed of the wave crest (and the rest of the wave) is $v = \frac{\lambda}{T} = f\lambda$,

where $f = \frac{1}{T}$ is the *frequency* in Hertz (Hz) of the wave (and of the vertical oscillation of the square point in the figure). One Hz is one cycle per second.

The mathematical form of the wave is $y = A\sin\left(2\pi\frac{x}{\lambda} - 2\pi\frac{t}{T}\right)$

Other forms for this equation include :

- $y = A\cos\left(2\pi\frac{x}{\lambda} - 2\pi\frac{t}{T}\right)$.
- $y = A\cos(kx - \omega t)$ *where* $k = \frac{2\pi}{\lambda}$ is called the wave number, and $\omega = \frac{2\pi}{T}$ is the angular frequency in radians per second.
- $y = A\cos(kx - 2\pi ft)$

If the wave travels to the left, then all the wave form equations above have the – sign replaced with a + sign.

At a given value of x (such as x=0), the equation for y is the equation of simple harmonic motion.

Traveling waves can carry energy from one place to another. The wave from some source can do work on a distant object, even though no material moves from the source to the distant object.

For a traveling sinusoidal wave, each bit of material executes simple harmonic motion. Each bit reaches its maximum displacement (amplitude) at a slightly different time from neighboring bits. As the wave crest travels to the right, a particular bit reaches maximum just a little later than the bit on its left, and just a little before the bit on its right.

PROPERTIES OF STANDING WAVES

The most commonly considered standing waves involve simple harmonic motion of an extended object, such as a taut string. In the following figure, imagine that the taut string is (somehow) free at the left end and fixed at the right end.

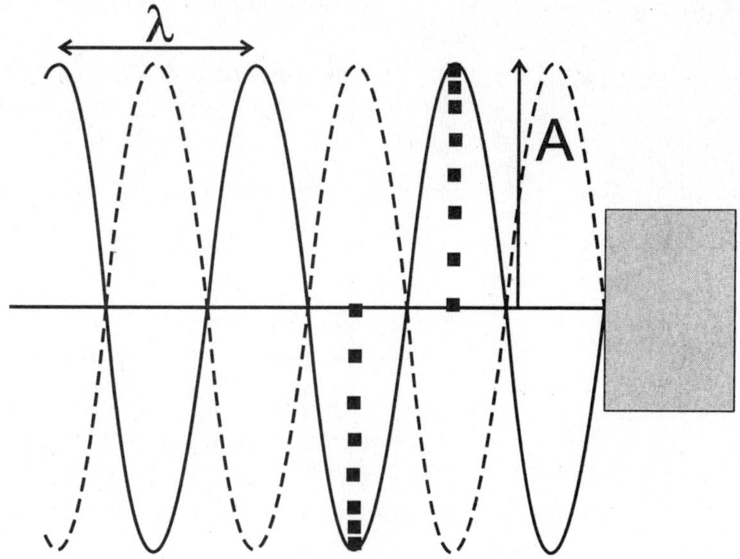

In contrast to the traveling wave, each bit of material reaches a maximum distance from equilibrium at the same time.

The solid line shows the distortion of the string at one time, and the dashed line shows the distortion of the string $\frac{1}{2}$ period later. After half a period, bits which had been displaced upward are now displaced downward and vice versa.

The right end of the string is tied to a block and cannot move. The left end of the string is free to move.

Some points on the string do not move at all. Those points are located in the figure above as the points where the dashed line crosses the solid line. These immobile points are called *nodes.*

Halfway between neighboring nodes are located the *antinodes,* points at which the string experiences its largest displacements.

The distance between maxima at any particular instant is equal to the wavelength of the standing wave. The distance between antinodes (and nodes) is only $\frac{1}{2}$ wavelength.

Note that for a string to exhibit a standing wave, its length must be specially related to the wavelength. Note that the string is $2\frac{3}{4}$ wavelengths long.

If both ends of the string were fixed, the string would have to be a whole number of half-wavelengths long. The same is true if both ends were free.

If one end is fixed and one end is free, the string must be an odd number of quarter-wavelengths long to exhibit a string wave.

The traveling wave equation $v = f\lambda$ continues to hold for the standing wave.

Because the boundaries allow only oscillations whose wavelengths "fit" the length of the string, most wavelengths are forbidden. This means that most oscillation frequencies, given by $f = \frac{v}{\lambda}$ are forbidden.

- For both ends fixed (or both free), the allowed frequencies of a string of length L are $f_n = n\frac{v}{L/2}$, where n is an integer.
- If one end is fixed and one free, the allowed frequencies are $f_n = (2n-1)\frac{v}{L/4}$. This formula ensures that only odd multiples of a quarter wavelength are allowed to stand on the string.

The speed of the wave for the string is $v = \sqrt{\frac{T}{\mu}}$, where T is now the *tension* in the string and μ is its mass per unit length. The standing wave may be represented in equation form:

- $y = A\sin(\omega t)\cos(kx)$ or
- $y = A\cos(\omega t)\sin(kx)$
- Use the boundary conditions to select the most convenient form. For example, if the displacement is fixed at x=0, then the second equation is convenient.

DOPPLER EFFECT

The *Doppler effect* is exemplified by the change in pitch of sound waves, caused either by motion of the source of the sound or motion of the detector of the sound. Velocities of both source and detector are measured relative to the air. (We may imagine that the air is always stationary.)

- In this discussion, velocities are positive if they are in the same direction as the direction of sound travel.
- A detector has positive velocity if it is moving away from the source.

- A source has positive velocity if it is moving towards the detector.

The source emits a steady tone of frequency $f_0 = \frac{1}{T_0}$. T_0 is the period of the sound source oscillation.

If the source is moving towards the detector with velocity v_S, then the separation between wave crests is reduced by the motion of the source. The wavelength becomes $\lambda = \lambda_0\left(1-\frac{v_S}{c}\right)$, where λ_0 is the wavelength observed when the source is stationary in the air, and *c* is speed of sound in the air. Because of the shorter wavelength, the detector observes a higher frequency, given by $f = f_0 \frac{1}{1-\frac{v_S}{c}}$.

If the detector is moving away from the source with velocity v_D, the time between arrivals of crests is increased because the detector is "running away." The observed period becomes $T = T_0 \frac{1}{1-\frac{v_D}{c}}$ and the observed frequency becomes $f = f_0\left(1-\frac{v_D}{c}\right)$.

Allowing for motion of both source and detector, we can write

$$f = f_0\left(\frac{1-\frac{v_D}{c}}{1-\frac{v_S}{c}}\right) = f_0\left(\frac{c-v_D}{c-v_S}\right).$$

There are, unfortunately, many ways to express this relation. In the version above, remember that positive velocities are in the direction of sound travel. This means that if the detector runs away from the source, the pitch goes down. If the source runs toward the detector, the pitch goes up.

SUPERPOSITION

When two waves arrive at the same place at the same time, they combine in the simplest possible way: their displacements simply add. This kind of combination is called *superposition.*

Since displacements can be positive or negative, the result of addition may be either smaller or larger than either of the original wave displacements.

- If the amplitude of a wave is decreased when a second wave is added to it, the superposition is said to be destructive. In this case the second wave is said to *interfere* with the first. It is said that the waves exhibit *destructive interference.*

- If the amplitude of a wave is increased when a second wave is added to it, the superposition is said to be constructive. It is said (somewhat incongruously) that the waves exhibit *constructive interference.*

- When two waves combine over an extended region of space, it is possible to have constructive interference in one place and destructive interference in another. The overall behavior is said to produce an *interference pattern.*

When two sound waves of the same frequency but from different loudspeakers arrive at the ear, the sound may be loud or soft, depending on the path taken by the two waves.

- If the paths are the same length, the two waves arrive with crests at the same time. They are said to arrive *in phase*. The same is true if the path lengths differ by a whole number of wavelengths.

- If the path lengths differ by an odd number of half-wavelengths, the interference is destructive. The sound is not loud.

- Moving the speakers or moving the ear changes the paths to the ear for the two waves, and one can hear changes from loud to soft and back to loud as the motion proceeds.

When two sound waves from the same speaker, but with different frequencies, arrive at the ear, they are sometimes in phase and sometimes "out of phase," interfering destructively. The ear hears a *flutter* or *beat* in the sound. The frequency of the perceived beat is equal to the difference in frequency between the two sound waves.

Standing waves on a string may be generated mathematically by the superposition of two identical waves traveling in opposite directions. The trigonometric identities for sums of angles show that

$\sin(\omega t - kx) + \sin(\omega t + kx) = 2\sin(\omega t)\cos(kx)$ and

$-\sin(\omega t - kx) + \sin(\omega t + kx) = 2\cos(\omega t)\sin(kx)$. The form that matches the boundary conditions at x=0 is the form to be chosen.

Chapter 15
PHYSICAL OPTICS

INTERFERENCE AND DIFFRACTION

- Light is a wave with very high frequency, very high speed, and very short wavelengths.
- In practice, it is difficult to demonstrate beats between light waves of two different frequencies.
- In practice, the wave-like properties that are demonstrated all have to do with waves that have identical frequencies but arrive following different paths to the detector.
- The speed of light in a vacuum is $c = 3\text{x}10^8 \dfrac{meters}{second}$.
- Light frequencies are on the order of $f = 6\text{x}10^{16}\, Hz$.
- Light wavelengths are on the order of $\lambda = 5\text{x}10^{-7}\, meters = 500nm$.

When light from a single source is sent through two slits, as in the following figure, the pattern of constructive and destructive interference has intensity maxima at angles given by $n\lambda = d \sin \theta_{BRIGHT}$, where d is the separation of the slits.

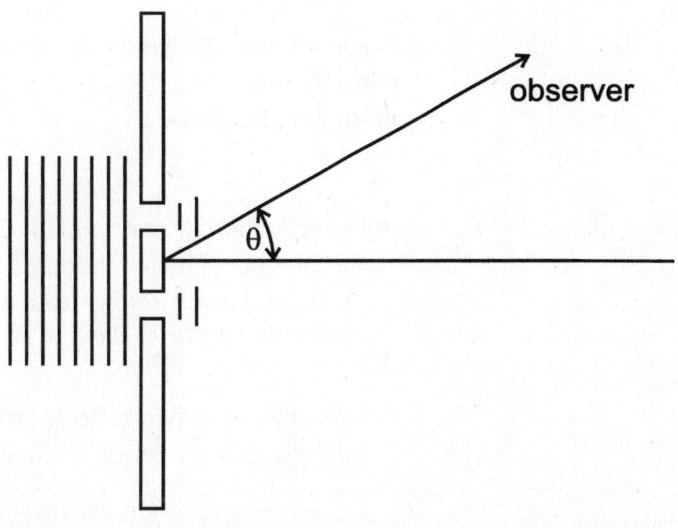

- This interference pattern is called a two-slit *diffraction pattern.*

- The same equation holds for the bright directions when the two slits are replaced by multiple slits, all separated by a distance, *d.*

- A multiple-slit device is called a *diffraction grating.* When light is passed through the grating, the bright light is concentrated much more at the exact angle of maximum intensity.

If the beam of light passes through a single slit, light radiating from different parts of the slit will also interfere constructively and destructively. Referring to the figure below, destructive interference produces dark spots at an angle given by $n\lambda = w\sin\theta_{DARK}$, where *w* is the width of the slit.

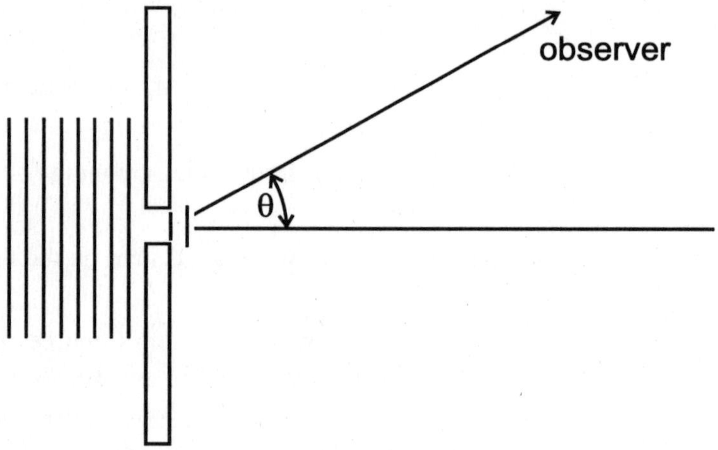

When a beam of light strikes a pair of parallel surfaces, each of which is partially reflecting, the reflected beam is the superposition of the results of multiple reflections.

- The reflection is *inverted* if it is caused by a surface of higher index of refraction. (The inversion of light reflected by a metal is an extreme case of this.) For reflection from a thin film of material, we may expect the reflection at one surface of the film to be inverted while the other is not.

- Consider only two reflections: the first reflection from the front surface and the first reflection from the back surface.

- If the excess path of the back surface reflection were a whole number of wavelengths, we would initially expect a bright reflection at that wavelength.

- Because one surface inverts the reflected light and the other does not, we see instead no reflection when the excess path is a whole number of wavelengths.

- For an oil film, the light is bright if twice the thickness of the film is an odd number of half wavelengths:

$$2\{thickness\} = \left(\frac{\lambda}{2}\right), 3\left(\frac{\lambda}{2}\right), 5\left(\frac{\lambda}{2}\right), 7\left(\frac{\lambda}{2}\right)\ldots$$

DISPERSION OF LIGHT AND THE ELECTROMAGNETIC SPECTRUM

Ordinary light does not come with a single wavelength. It is not *monochromatic*.

- Light from an incandescent lamp bulb has a continuous distribution of wavelengths. The visible portion of this distribution covers about 800 nm to about 400nm.

- Each wavelength produces its own color sensation in the human eye. 800 nm is deep red, and 400 is violet, or deep blue. In between, the wavelengths track with the colors of the rainbow. For example, light of wavelength 560 nm is green.

- Light from gas lamps (such as mercury vapor or low pressure sodium vapor) has a more restricted sampling of wavelengths.

Light is an electromagnetic wave. The range of wavelengths produced by a light source is called the *spectrum* of the source.

- The electromagnetic spectrum extends to wavelengths much longer and much shorter than its visible portion.

- Radio waves have wavelengths as long as many meters. Radar wavelengths are measured in centimeters, and infrared radiation is measured in micrometers.

- Ultraviolet light, x-rays, and gamma rays have wavelengths much shorter than visible light.

Re-consider the equation developed for a grating and a monochromatic light source of wavelength λ: $n\lambda = d \sin\theta_{BRIGHT}$.

- Imagine that we consider only $n = 1$ and allow multiple wavelengths in our light source.

- Each wavelength will be bright at its own angle, as determined by $\lambda = d \sin\theta_{BRIGHT}$.

- The grating *disperses* the light, showing how much intensity is present at each wavelength.

For incandescent lamps and for sunlight, the grating reproduces the same separation of colors (wavelengths) as a natural rainbow.

Chapter 16
GEOMETRIC OPTICS

REFLECTION AND REFRACTION

When a light wave encounters a good conductor, its electromagnetic character determines the outcome.

- The conductor does not allow electric fields within it and adjusts its mobile charges to make the internal field zero.
- This charge motion radiates a reflected light wave at the same frequency.

Experiment and theory (Huygens' Principle) agree that when a light beam strikes a flat surface, the reflected light

- reverses the component of velocity that is perpendicular ("normal") to the surface and
- does not change the component of velocity that is parallel to the surface.

This result is summarized by saying that, "The angle of incidence equals the angle of reflection." Traditionally, the angles are measured from the light beam to a line perpendicular to the surface. That line is called the "normal" to the surface.

When a light beam encounters a dielectric medium, its electromagnetic character determines the outcome. The dipoles in the dielectric medium move in response to the electric fields, reducing the electric field within the medium. Some of the light is reflected because of the charge motion. As with a metal, the angle of reflection equals the angle of incidence. Some of the light continues to propagate through the dielectric medium. The frequency is the same as in a vacuum, but the speed of the wave is reduced.

Experiment and theory (Huygens' Principle) agree that the light changes direction when it enters the new medium. Rays that are bent because they move from one medium to another, are said to be *refracted.* Light rays in the medium with higher index of refraction (lower light speed) are oriented more towards the normal.

Snell's law describes the way that the direction changes: When light passes from medium 1 to medium 2, the angles to the normal (theta) of the ray in each medium are related by $n_1 \sin\theta_1 = n_2 \sin\theta_2$. n is called the index of refraction of the medium. The index of refraction is the ratio of the speed of light in the vacuum to the speed of light in the medium. It is 1 for a vacuum, and greater than 1 (typically less than 2) for all material mediums. The slower velocity in a dielectric medium makes the light wavelength shorter, for the same frequency. The short wavelength is calculated by dividing the vacuum wavelength by the index of refraction. When an observer looks down into a fish tank at a rock on the bottom, refraction bends the rays coming from the water to the air (and thence the eye) so that the rock looks closer to the surface than it really is.

MIRRORS

Mirrors are used to create optical images of real objects. For a flat mirror, the image is the same size as the object and is located behind the surface, a distance equal to the distance from the object to the front of the mirror.

- This image is said to be *virtual* because no light actually passes through its location.
- This result may be proved using the law of reflection (angle of incidence equals angle of reflection) and ray diagrams.

CONCAVE SPHERICAL MIRRORS

The behavior of a *concave spherical mirror* can be deduced from the way that it reflects a beam of parallel light rays. The rays are directed along the optical axis of the mirror.

As long as the diameter of the mirror is small compared to its radius of curvature, the parallel rays are all focused to a point, the *focal point.* For a spherical mirror, the focal point is half way between the mirror and its center of curvature.

The distance from the mirror to the focal point is called the focal length, and given the symbol, *f.*

When light from a real object strikes the curved mirror,

- some of the rays arrive parallel to the optic axis and are reflected through the focal point.
- some arrive by passing through the focus and are reflected parallel to the optic axis.

- the rays cross at the location of the image formed by the mirror.

The location of the image is calculated from $\frac{1}{D_O} + \frac{1}{D_I} = \frac{1}{f}$, where D_O is the distance from the object to the mirror, D_I is the distance from the image to the mirror, and f is the focal length. The height of the image, h_I, is given by $h_I = -\left(\frac{D_I}{D_O}\right)h_O$, where h_O is the height of the object. The minus sign indicates that if both *Ds* are positive, the image is inverted. If $D_O < f$, D_I will be negative. This indicates that the image is *virtual,* located behind the mirror. The formula for the height continues to work, indicating that now the image is erect.

CONVEX SPHERICAL MIRRORS

The behavior of a *convex spherical mirror* can be deduced from the way that it reflects a beam of parallel light rays. The rays are directed along the optical axis of the mirror.

As long as the diameter of the mirror is small compared to its radius of curvature, the parallel rays reflect as if from a focal point located behind the mirror. This mirror is said to have a virtual focus. As before, the focal length is half the radius of curvature, but this time it is negative. When light from a real object strikes the curved mirror,

- some of the rays arrive parallel to the optic axis, and are reflected as if they originated at the virtual focal point.

- some of the rays arrive along a line that would cross the focal point. They are reflected before they reach the virtual focus and travel away parallel to the optic axis.

- the rays do not cross. However, lines extended back through the mirror surface do cross at the location of the image formed by the mirror.

- the image is virtual. No light passes through it.

The location of the image is calculated as before, using $\frac{1}{D_O}+\frac{1}{D_I}=\frac{1}{f}$,

where D_O is the distance from the object to the mirror, D_I is the distance from the image to the mirror, and f is the focal length.

- D_O is positive in our applications.
- f is negative for the convex mirror
- D_I is negative for our applications. The image is formed behind the mirror surface. It is virtual and erect.

The height of the image, h_I, is given by $h_I=-\left(\frac{D_I}{D_O}\right)h_O$, where h_O is the height of the object. The minus sign indicates that if both *Ds* are positive, the image is inverted. In practice, for the convex mirror, the sign is plus and the image is erect. If the object moves to make $D_O<|f|$, no particular change occurs for the convex mirror.

LENSES

When glass or plastic presents a curved surface to a beam of parallel light rays, the rays strike the surface at a variety of angles. As a result, they are refracted at a variety of angles.

If the surface is convex, the rays, bent towards the local normal, converge on a focus within the material.

Generally, the light exits the other side of the glass before reaching a focus. The curvature of the second glass surface modifies the location of the focus.

- The two-surface object is a lens.
- The lens maker's formula for the focal length of the lens is $\frac{1}{f}=(n-1)\left(\frac{1}{R_1}-\frac{1}{R_2}\right)$.
- R_1 is the radius of curvature of the first surface (the one the light hits first). R_1 is positive if that surface bulges towards the incoming light.

- R_2 is the radius of curvature of the second surface (the one the light hits second). R_2 is positive if that surface bulges towards the incoming light.

- The focal length is the same on both sides of the lens.

If the focal length of the lens is positive, it is called a *converging lens*. It focuses parallel rays to a real focal point.

If the focal length is negative, the lens is called a *diverging lens*. Parallel light passing into the lens diverges as though it came from a point on the "incoming" side of the lens. This lens has a virtual focus.

The location and size of the image are calculated with the same formula that worked for the mirror: $\frac{1}{D_O}+\frac{1}{D_I}=\frac{1}{f}$ and $h_I=-\left(\frac{D_I}{D_O}\right)h_O$.

When the object sends light through the lens to form a real image on the other side, both the object distance and the image distance are positive. This real image will be inverted.

As before, negative image distance means the image is virtual and erect. This means that the image is on the same side of the lens as the object.

A lens used as a magnifier is placed so that the object is just inside the focal length. The lens produces a virtual erect image farther from the lens than the object. This allows the viewer to hold the object closer to the eye, making the image on the retina larger. The view of the object is magnified.

Lens Combinations

The magnifier can also be used to magnify a real image.

- If the real image is formed by a long focal length lens from a distant object, the lens combination is a telescope. The purpose of the long focal length lens is to gather light to make a bright image for the magnifier to work on.

- If the real image is formed by a short focal length lens from a nearby object, the combination is called a microscope.

If two lenses are placed side-by-side, the combination forms an image as if it were a single lens with an effective focal length given by

$$\frac{1}{f_1}+\frac{1}{f_2}=\frac{1}{f_{EFFECTIVE}}.$$

- Because the inverses of the focal lengths appear in lens combinations, a new quantity called the power of the lens is defined as $\frac{1}{f}$.
- If the focal length is 1 meter, the lens is said to have a power of 1 diopter. A lens with a focal length of 20 cm has a power of 5 diopters.

UNIT 5
Atomic Physics and Quantum Effects

Chapter 17

ATOMIC PHYSICS AND QUANTUM EFFECTS

ALPHA PARTICLE SCATTERING AND THE RUTHERFORD MODEL

Alpha particles are emitted by radioactive materials. They consist of two neutrons and two protons bound together. The nucleus of ordinary helium is an alpha particle. The charge is positive because there are two protons. The mass of an alpha particle is about 7,000 times that of an electron.

In the 19th century, the atomic model of matter was not universally accepted. The conservative model of matter was that it was more or less uniformly distributed in space, even on a microscopic scale. There was very little experimental evidence to the contrary.

Geiger and Marsden placed a gold foil in a beam of alpha particles. From the reduction in the beam intensity they could estimate the strength of the interaction between the alpha particles and the gold.

They also determined what happened to the alpha particles that were removed from the beam by the gold. Some just stopped, but others were scattered out of the beam, leaving the gold in different directions.

The surprising result was that some alpha particles bounced straight back. The only tenable explanation was that those alpha particles had collided with a particle in the gold and that the particle was much more massive than the alpha particle.

Rutherford derived a formula for the distribution of scattered alpha particles, based on a model in which the positive charges in matter are concentrated in atomic nuclei. The formula fit the observations and gave the first strong evidence in favor of the Rutherford model of the atom.

As the model developed, the atom was pictured like a tiny solar system, with the positive nucleus at the center and the small electrons orbiting like planets. In this "solar system," the attractive force was the electrical force rather than the gravitational force.

PHOTONS AND THE PHOTOELECTRIC EFFECT

The *photoelectric* effect was known in the 19th century but not understood. Two electrodes are enclosed in a vacuum with leads leaving the walls of the vacuum chamber. Light is shined onto one of the electrodes. It is observed that this electrode becomes positive while the other becomes negative. When the two electrodes are connected, a current is observed to flow between them.

Electrons leave the illuminated electrode, migrate to the other electrode, and then return "home" via the external wire connecting the two electrodes.

The energy of the ejected electrons was found not to depend on the intensity of the light. The current was proportional to the light intensity.

That is, more intense light liberated more electrons.
Einstein used an idea of Planck to propose a particle model for light: In the model, light consists of photons.

- Each photon has an energy $E = hf$,where f is the frequency of the light, and h is a new constant called *Planck's constant.*

- Since $h = 6.6\text{x}10^{-34}\ Joule\ \sec$ and $f \cong 6\text{x}10^{14}\ Hz$, the energy of a single photon is very small, on the order of $4x10^{-19}\ Joule$.

- This energy is often reexpressed in units of electron volts, eV: $1eV = 1.6\text{x}10^{-19}\ Joule$. Then, a photon has an energy of a few electron volts.

- Photons have zero mass and travel at the speed of light.

- Einstein predicted that the energy of the ejected electrons would be proportional to the light frequency, with slope equal to Planck's constant. The prediction was verified by Millikan, and Einstein won the Nobel prize for this work.

- The prediction may be summarized with an equation for the kinetic energy of the "photoelectron" (the electron ejected from the electrode by an incoming photon): $KE = hf - W$, where W is called the *work function*. W is the work that must be done to unbind the electron from its metallic home. At low frequencies, the photon is not energetic enough to do this work, and no electrons are ejected.

THE BOHR MODEL AND ENERGY LEVELS

The diffraction gratings discussed in Chapter 15, *Physical Optics*, disperse light according to wavelength so that the experimenter can observe what wavelengths are present in light from any source.

Low pressure electric discharge gas lamps, such as those made from neon, mercury, and sodium, are found to lack many of the wavelengths that make up the "rainbow" spectrum of an incandescent lamp.

Gas discharge lamps emit only a countable number of discrete wavelengths. The Bohr model of the atom was invented to account for this extreme limitation on allowed wavelengths.

Einstein's photons showed that if only a few wavelengths are emitted, then the photons that are emitted come in only a few different energies.

Recall that $E_{PHOTON} = hf = h\frac{c}{\lambda}$.

Conservation of energy and work by Planck on photon emission, led to the idea that the electrons on an atom must exist only in a limited number of energy states.

- When the electron emits a photon, it lowers its energy by moving from one state to a state of lower energy.

- If there are only a few such states, then only a few energies (and thus wavelengths) are allowed to the emitted photons.

- The energy of the atomic electrons is said to be *quantized,* meaning that electron energies must be chosen from a discrete list of values.

Bohr adopted the idea that the angular momentum, *L,* of an orbiting electron is quantized: $L = n\frac{h}{2\pi} = n\hbar$, where *n* is an integer. Application of this quantization to a single electron orbiting a single proton (the hydrogen atom) leads to allowed energy states given by

$E_n = -\frac{me^4}{(8)(n^2)\varepsilon_0^2 h^2} = \frac{E_1}{(n^2)}$, where *m* is the mass of the electron.

- For hydrogen, the ground state energy E_1 is

  $E_1 = -13.6eV = -2.2\text{x}10^{-18}\,Joule$.

- In this equation, if the electron energy is greater than zero, it is free—not bound to the proton.

- Energy differences between states with *n*=1,2,3, etc., give excellent agreement with the photon energies measured for the light emitted by hydrogen.

These allowed states had specific allowed circular orbit radii, given by

$r_n = -\frac{(n^2)h^2\varepsilon_0}{\pi me^2} = (n^2)a_0$. a_0 is called the radius of the first Bohr orbit.

WAVE-PARTICLE DUALITY

Evidence has accumulated over the past 100 years that an electron can act as a wave or as a particle, depending on the experiment in which it participates. Electron diffraction and interference have been observed, and yet a television picture tube successfully treats electrons as particles.

One connection between the wave character and the particle character is in the relation of momentum to wavelength: $\lambda = \frac{h}{p}$. In this equation, Planck's constant (divided by 2π) and the momentum, $p=mv$ are well defined. The wavelength, λ, when used in diffraction formulas, correctly predicts the pattern of diffraction of a beam of electrons. The wavelength is called the *DeBroglie wavelength*, after the man who proposed its existence

Another connection is the *uncertainty principle*. It is concerned with measurements of more than one property of a particle. The most straightforward example is the measurement of the momentum and position of a particle. In one dimension, the rule is $\Delta p \Delta x \geq \hbar$.

- Δp and Δx are to be interpreted as the uncertainties in the measurement of momentum and position, repectively.

- The rule says that if the experiment is designed to give a very small uncertainty in momentum, then the position cannot be well known. Another way to say this is that it takes a large distance to obtain a very accurate measurement of momentum.

- Classical physics did not have this limitation on accuracy. The only limit was thought to be the ability of the experimenter.

- Experiments have shown that the uncertainty principle is correct.

DeBroglie found that he could derive the entire Bohr quantization scheme by requiring that an electron orbit contain an integral number of electron wavelengths along its circumference. This was the earliest strong indication that the wave character of particles was a real property.

Chapter 18
NUCLEAR PHYSICS

RADIOACTIVITY AND HALF-LIFE

Some chemical substances, called *radioactive*, emit particles and transmute into different chemical substances. The process takes place one atom at a time. The transmutation of the nucleus is by far more energetic than the concurrent change in the electron orbits.

The process is called *nuclear decay,* because the new atoms (the *daughter nuclei*) have smaller nuclei than the original atoms (the *parent nuclei*).

Often, the emitted particles are also nuclei. In this case, each decay can have more than one daughter particle. If the two daughter particles are of roughly the same size, the process is called fission.

Experimentally, it is impossible to predict when a particular nucleus will decay. It is only possible to give a probability that it will decay in the next second. If a large number of nuclei are present, this makes it possible to predict how many will decay in the next second. That number is called the *rate of decay* of the sample of radioactive material. The decay rate is proportional to the number of radioactive parent particles in the sample.

$RATE = \dfrac{dN}{dt} = -\lambda t$, where N is the number of parent nuclei in the sample.

λ is called the decay constant. It is different for each different kind of parent nucleus. This rate can be integrated to give N as a function of time: $N = N_0 e^{-\lambda t}$. The time for N to fall to half of its beginning value is called the *half-life*, $T_{1/2} = \dfrac{\ln(2)}{\lambda}$.

NUCLEAR REACTIONS

In a nuclear reaction, the transmutation of a nucleus is caused by its interaction with another particle. Two numbers are important in nuclear reactions:

- The number of protons determines the chemical species of the nucleus by determining the number of electrons on the atom. This number can change in a nuclear reaction. It is called the *atomic number* and is given the symbol *Z*.

- The *mass number*, symbolized by A, is the sum of the number of neutrons and the number of protons.

The mass number is conserved in nuclear reactions. That is, the number of neutrons and the number of protons present may change, but the sum of all nucleons (neutrons plus protons) does not change. These numbers are given in the form ${}^{A}_{Z}X$ for a chemical element named X. With the exception of hydrogen, A is always greater than Z.

For a given chemical species, there is only one value for Z, but several values for A are allowed. This is because the number of neutrons can vary without changing species.

- Typically, the number of neutrons is equal to or greater than Z.

- Nuclei with the same Z but different A are called isotopes. The chemical character of isotopes of the same species is the same for all isotopes, but the masses are all different.

- Some isotopes are stable, while others are not. When the nucleus of an unstable isotope decays, it may or may not decay to a new species, with new (lower) Z.

In nuclear reactions, charge is conserved in the following sense: the total charge of all the protons and electrons remains constant. In some reactions, a proton and an electron may disappear, while a neutron appears. The total charge remains constant.

When a parent nucleus decays, it releases energy and decays to a system which has lower energy than the parent.

- In a very simple example, the nucleus emits a gamma ray and settles into a more strongly bound state.

- It is found experimentally that, after the decay and emission of energy, the total mass of the system (including all the masses of all decay products) is smaller.

- Calculation shows that the lost mass, Δm, is related to the energy released by the decay by the famous Einstein equation, $E_{RELEASED} = (\Delta m)c^2$.

- Similar changes in mass occur when a chemical reaction occurs, but the change is very small because the energy released is much smaller than the nuclear energy released.

The most commercial nuclear reaction is the uranium fission reaction:

$^{235}_{92}U$ absorbs a neutron and then (typically) breaks into two daughters plus neutrons.

- Many daughters are possible, but they are of similar size.
- When the fragments are nearly equal, the process is called fission, because it resembles the fission of a biological cell.
- Most of the energy released appears as kinetic energy of the fragments, and eventually as heat, as they are brought to rest.
- Then neutrons are then available for capture by other $^{235}_{92}U$ nuclei. If this happens, the next uranium nucleus can repeat the process, as part of a chain reaction.

Typically, daughter nuclei are also radioactive, and a chain of radio active decays ensues, lasting until it ends in a non-radioactive nucleus.

The most strongly bound, lowest energy nuclei are in the neighborhood of iron on the periodic table. This fact means that lighter nuclei can release energy by fusing together and making themselves into nuclei nearer in mass to iron. Once again the Einstein equation relates the mass lost to the energy released. Fusion has been used for bombs, but not for commercial power generation.

ANSWER SHEET FOR PHYSICS B PRACTICE TEST I

Section I: Multiple Choice

1 ⊂A⊃ ⊂B⊃ ⊂C⊃ ⊂D⊃ ⊂E⊃
2 ⊂A⊃ ⊂B⊃ ⊂C⊃ ⊂D⊃ ⊂E⊃
3 ⊂A⊃ ⊂B⊃ ⊂C⊃ ⊂D⊃ ⊂E⊃
4 ⊂A⊃ ⊂B⊃ ⊂C⊃ ⊂D⊃ ⊂E⊃
5 ⊂A⊃ ⊂B⊃ ⊂C⊃ ⊂D⊃ ⊂E⊃
6 ⊂A⊃ ⊂B⊃ ⊂C⊃ ⊂D⊃ ⊂E⊃
7 ⊂A⊃ ⊂B⊃ ⊂C⊃ ⊂D⊃ ⊂E⊃
8 ⊂A⊃ ⊂B⊃ ⊂C⊃ ⊂D⊃ ⊂E⊃
9 ⊂A⊃ ⊂B⊃ ⊂C⊃ ⊂D⊃ ⊂E⊃
10 ⊂A⊃ ⊂B⊃ ⊂C⊃ ⊂D⊃ ⊂E⊃
11 ⊂A⊃ ⊂B⊃ ⊂C⊃ ⊂D⊃ ⊂E⊃
12 ⊂A⊃ ⊂B⊃ ⊂C⊃ ⊂D⊃ ⊂E⊃
13 ⊂A⊃ ⊂B⊃ ⊂C⊃ ⊂D⊃ ⊂E⊃
14 ⊂A⊃ ⊂B⊃ ⊂C⊃ ⊂D⊃ ⊂E⊃
15 ⊂A⊃ ⊂B⊃ ⊂C⊃ ⊂D⊃ ⊂E⊃
16 ⊂A⊃ ⊂B⊃ ⊂C⊃ ⊂D⊃ ⊂E⊃
17 ⊂A⊃ ⊂B⊃ ⊂C⊃ ⊂D⊃ ⊂E⊃
18 ⊂A⊃ ⊂B⊃ ⊂C⊃ ⊂D⊃ ⊂E⊃
19 ⊂A⊃ ⊂B⊃ ⊂C⊃ ⊂D⊃ ⊂E⊃
20 ⊂A⊃ ⊂B⊃ ⊂C⊃ ⊂D⊃ ⊂E⊃
21 ⊂A⊃ ⊂B⊃ ⊂C⊃ ⊂D⊃ ⊂E⊃
22 ⊂A⊃ ⊂B⊃ ⊂C⊃ ⊂D⊃ ⊂E⊃
23 ⊂A⊃ ⊂B⊃ ⊂C⊃ ⊂D⊃ ⊂E⊃
24 ⊂A⊃ ⊂B⊃ ⊂C⊃ ⊂D⊃ ⊂E⊃
25 ⊂A⊃ ⊂B⊃ ⊂C⊃ ⊂D⊃ ⊂E⊃
26 ⊂A⊃ ⊂B⊃ ⊂C⊃ ⊂D⊃ ⊂E⊃
27 ⊂A⊃ ⊂B⊃ ⊂C⊃ ⊂D⊃ ⊂E⊃
28 ⊂A⊃ ⊂B⊃ ⊂C⊃ ⊂D⊃ ⊂E⊃
29 ⊂A⊃ ⊂B⊃ ⊂C⊃ ⊂D⊃ ⊂E⊃
30 ⊂A⊃ ⊂B⊃ ⊂C⊃ ⊂D⊃ ⊂E⊃
31 ⊂A⊃ ⊂B⊃ ⊂C⊃ ⊂D⊃ ⊂E⊃
32 ⊂A⊃ ⊂B⊃ ⊂C⊃ ⊂D⊃ ⊂E⊃
33 ⊂A⊃ ⊂B⊃ ⊂C⊃ ⊂D⊃ ⊂E⊃
34 ⊂A⊃ ⊂B⊃ ⊂C⊃ ⊂D⊃ ⊂E⊃
35 ⊂A⊃ ⊂B⊃ ⊂C⊃ ⊂D⊃ ⊂E⊃
36 ⊂A⊃ ⊂B⊃ ⊂C⊃ ⊂D⊃ ⊂E⊃
37 ⊂A⊃ ⊂B⊃ ⊂C⊃ ⊂D⊃ ⊂E⊃
38 ⊂A⊃ ⊂B⊃ ⊂C⊃ ⊂D⊃ ⊂E⊃
39 ⊂A⊃ ⊂B⊃ ⊂C⊃ ⊂D⊃ ⊂E⊃
40 ⊂A⊃ ⊂B⊃ ⊂C⊃ ⊂D⊃ ⊂E⊃
41 ⊂A⊃ ⊂B⊃ ⊂C⊃ ⊂D⊃ ⊂E⊃
42 ⊂A⊃ ⊂B⊃ ⊂C⊃ ⊂D⊃ ⊂E⊃
43 ⊂A⊃ ⊂B⊃ ⊂C⊃ ⊂D⊃ ⊂E⊃
44 ⊂A⊃ ⊂B⊃ ⊂C⊃ ⊂D⊃ ⊂E⊃
45 ⊂A⊃ ⊂B⊃ ⊂C⊃ ⊂D⊃ ⊂E⊃
46 ⊂A⊃ ⊂B⊃ ⊂C⊃ ⊂D⊃ ⊂E⊃
47 ⊂A⊃ ⊂B⊃ ⊂C⊃ ⊂D⊃ ⊂E⊃
48 ⊂A⊃ ⊂B⊃ ⊂C⊃ ⊂D⊃ ⊂E⊃
49 ⊂A⊃ ⊂B⊃ ⊂C⊃ ⊂D⊃ ⊂E⊃
50 ⊂A⊃ ⊂B⊃ ⊂C⊃ ⊂D⊃ ⊂E⊃
51 ⊂A⊃ ⊂B⊃ ⊂C⊃ ⊂D⊃ ⊂E⊃
52 ⊂A⊃ ⊂B⊃ ⊂C⊃ ⊂D⊃ ⊂E⊃
53 ⊂A⊃ ⊂B⊃ ⊂C⊃ ⊂D⊃ ⊂E⊃
54 ⊂A⊃ ⊂B⊃ ⊂C⊃ ⊂D⊃ ⊂E⊃
55 ⊂A⊃ ⊂B⊃ ⊂C⊃ ⊂D⊃ ⊂E⊃
56 ⊂A⊃ ⊂B⊃ ⊂C⊃ ⊂D⊃ ⊂E⊃
57 ⊂A⊃ ⊂B⊃ ⊂C⊃ ⊂D⊃ ⊂E⊃
58 ⊂A⊃ ⊂B⊃ ⊂C⊃ ⊂D⊃ ⊂E⊃
59 ⊂A⊃ ⊂B⊃ ⊂C⊃ ⊂D⊃ ⊂E⊃
60 ⊂A⊃ ⊂B⊃ ⊂C⊃ ⊂D⊃ ⊂E⊃
61 ⊂A⊃ ⊂B⊃ ⊂C⊃ ⊂D⊃ ⊂E⊃
62 ⊂A⊃ ⊂B⊃ ⊂C⊃ ⊂D⊃ ⊂E⊃
63 ⊂A⊃ ⊂B⊃ ⊂C⊃ ⊂D⊃ ⊂E⊃
64 ⊂A⊃ ⊂B⊃ ⊂C⊃ ⊂D⊃ ⊂E⊃
65 ⊂A⊃ ⊂B⊃ ⊂C⊃ ⊂D⊃ ⊂E⊃
66 ⊂A⊃ ⊂B⊃ ⊂C⊃ ⊂D⊃ ⊂E⊃
67 ⊂A⊃ ⊂B⊃ ⊂C⊃ ⊂D⊃ ⊂E⊃
68 ⊂A⊃ ⊂B⊃ ⊂C⊃ ⊂D⊃ ⊂E⊃
69 ⊂A⊃ ⊂B⊃ ⊂C⊃ ⊂D⊃ ⊂E⊃
70 ⊂A⊃ ⊂B⊃ ⊂C⊃ ⊂D⊃ ⊂E⊃

Section I: Free Response

1 ⊂A⊃ ⊂B⊃ ⊂C⊃ ⊂D⊃ ⊂E⊃
2 ⊂A⊃ ⊂B⊃ ⊂C⊃ ⊂D⊃ ⊂E⊃
3 ⊂A⊃ ⊂B⊃ ⊂C⊃ ⊂D⊃ ⊂E⊃
4 ⊂A⊃ ⊂B⊃ ⊂C⊃ ⊂D⊃ ⊂E⊃
5 ⊂A⊃ ⊂B⊃ ⊂C⊃ ⊂D⊃ ⊂E⊃
6 ⊂A⊃ ⊂B⊃ ⊂C⊃ ⊂D⊃ ⊂E⊃

Physics Formulas

TABLE OF INFORMATION

Constants and Conversion Factors

1 unified atomic mass unit.	$1\ u = 1.66 \times 10^{-27}$ kg = 931 MeV / c^2
Proton mass	$m_p = 1.67 \times 10^{-27}$ kg
Neutron mass	$m_n = 1.67 \times 10^{-27}$ kg
Electron mass	$m_e = 9.11 \times 10^{-31}$ kg
Magnitude of electron charge	$e = 1.60 \times 10^{-19}$
Avogadro's number	$N_0 = 6.02 \times 10^{23}\ mol^{-1}$
Universal gas constant	$R = 8.31$ J / (mol × K)
Boltzmann's constant	$k_B = 1.38 \times 10^{-23}$ J/K
Speed of light	$c = 3.00 \times 10^8$ m/s
Planck's constant	$h = 6.63 \times 10^{-34} J \times s = 4.14 \times 10^{-15}\ eV \times s$
	$Hc = 1.99 \times 10^{-25} J \times m = 1.24 \times 10^3\ eV \times nm$
Vacuum permittivity	$e_0 = 8.85 \times 10^{-12} C2/N \times m^2$
Coulomb's law constant	$k = 1/4\pi e_0 = 9.0 \times 10^9 N \times m^2/C^2$
Vacuum permeability	$m_0 = 4p \times 10^{-7} Wb/(A \times m)$
Magnetic constant	$k' = m_0/4\pi \times 10^{-7} Wb/(A \times m)$
Universal gravitational constant	$G = 6.67 \times 10^{-11}\ m^3/kg \times s^2$
Acceleration due to gravity at the Earth's surface	$g = 9.8\ m/s^2$
1 atmosphere pressure	$1\ atm = 1.0 \times 10^5\ N/m^2 = 1.0 \times 10^5$ Pa
1 electron volt	$1\ eV = 1.60 \times 10^{-19} J$
1 angstrom	$1\ Å = 1 \times 10^{-10}$ m

Units

Name	Symbol
meter	m
kilogram	kg
second	s
ampere	A
kelvin	K
mole	mol
hertz	Hz
newton	N
pascal	Pa
joule	J
watt	W
coulomb	C
volt	V
ohm	W
henry	H
farad	F
weber	Wb
tesla	T
degree Celsius	°C
electron-volt	eV

Prefixes

Factor	Prefix	Symbol
10^9	giga	G
10^6	mega	M
10^3	kilo	k
10^{-2}	centi	c
10^{-3}	milli	m
10^{-6}	micro	m
10^{-9}	nano	n
10^{-12}	pico	p

Values of Trigonometric Functions For Common Angles

Angle	Sin	Cos	Tan
0°	0	1	0
30°	1 / 2	$\sqrt{3}/2$	$\sqrt{3}/3$
37°	3 / 5	4 / 5	3 / 4
45°	$\sqrt{2}/2$	$\sqrt{2}/2$	1
53°	4 / 5	3 / 5	4 / 3
60°	$\sqrt{3}/2$	1 / 2	$\sqrt{3}$
90°	1	0	∞

Newtonian Mechanics

a = acceleration	F = force
f = frequency	h = height
J = impulse	K = kinetic energy
k = spring constant	l = length
m = mass	N = normal force
P = power	p = momentum
r = radius or distance	s = displacement
T = period	t = time
U = potential energy	v = velocity or speed
W = work	x = position

Physics B

PRACTICE TEST I

SECTION I—MULTIPLE CHOICE

Directions: Each question listed below has five possible choices. Select the best answer given the information in each problem and mark the corresponding oval on your answer sheet. (You may assume $g = 10 \text{ m/s}^2$).

1. When a propeller-driven airplane takes off, its nose lifts up. Due to the counter-clockwise motion of its propeller (as seen by the pilot), it will also tend to

    (A) tilt clockwise, as seen by the pilot.
    (B) tilt counterclockwise, as seen by the pilot.
    (C) tilt counterclockwise, as seen by the pilot, and turn left, as seen from above.
    (D) tilt clockwise, as seen by the pilot, and turn right, as seen from above.
    (E) tilt counterclockwise, as seen by the pilot, and turn right, as seen from above.

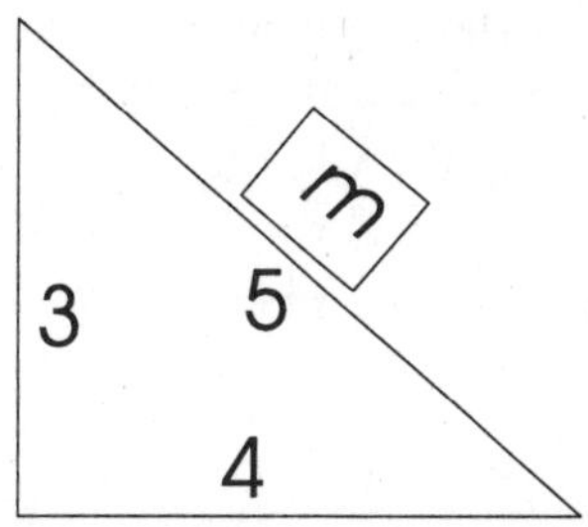

2. A block slides down an inclined plane, as shown above, at constant velocity. The coefficient of kinetic friction between the block and plane is

    (A) $\frac{3}{4}$.
    (B) $\frac{2}{3}$.
    (C) $\frac{4}{5}$.
    (D) $\frac{1}{2}$.
    (E) $\frac{3}{2}$.

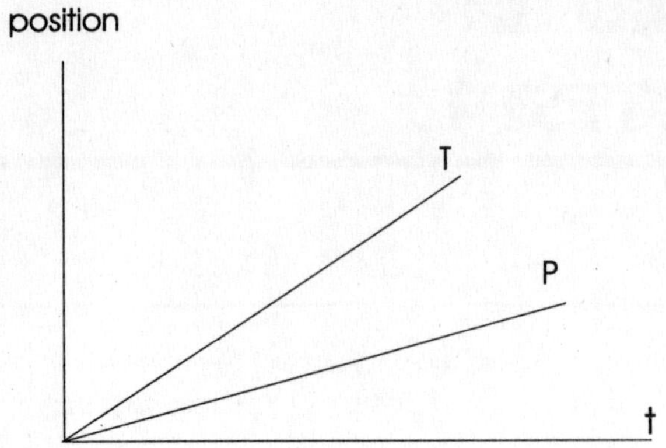

3. A stunt flyer in a small airplane is attempting to land on the back of a truck. The positions of the plane, *P*, and truck, *T*, are shown in the above graph. Assuming the velocity of the truck does not change, in order to land on the small truck and not crash, the plane must

(A) accelerate.
(B) decelerate.
(C) stay at constant velocity.
(D) accelerate, then decelerate.
(E) decelerate, then accelerate.

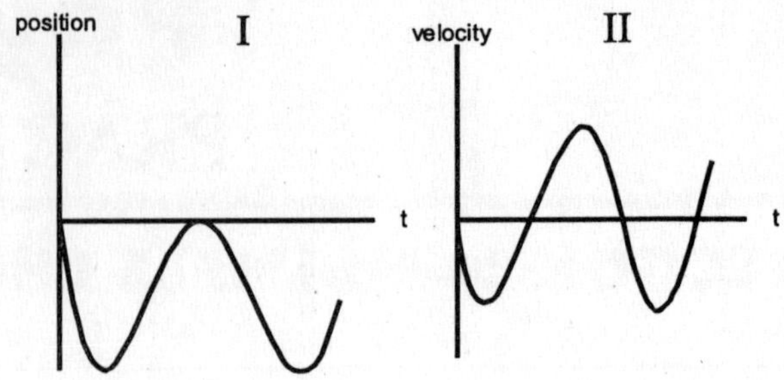

4. A bungee jumper jumps from a bridge (height $h = 0$) down into a deep ravine. Her position and velocity are correctly given by

(A) both the above graphs.
(B) graph I only.
(C) graph II only.
(D) both graphs, if down is changed to the positive direction.
(E) neither graph.

5. A boy runs and jumps horizontally off a dock $\frac{5}{4}$ m above the water and lands in the lake 2m away. His velocity at the end of the dock is most closely

(A) 1m/s.
(B) 2m/s.
(C) 3m/s.
(D) 4m/s.
(E) 5m/s.

6. A soldier hiding in a tree fires a rifle horizontally with a muzzle velocity of 700 m/s at another soldier hiding in a tree half a mile away. The soldier in the other tree simultaneously rolls off his branch and falls to the ground. The bullet

(A) hits the soldier, unless he is too close to the ground.
(B) hits the soldier.
(C) misses the soldier.
(D) hits the exact spot he left.
(E) misses the soldier, unless he is too close to the ground.

7. A helicopter flies with its nose pointed due west for four hours at 100 km/hr. The total distance to the airport it travels to is 500 km. The wind speed and direction could be

(A) east at 20 km/hr.
(B) west at 20 km/hr.
(C) north at 30 km/hr.
(D) south at 30 km/hr.
(E) north at 75 km/hr.

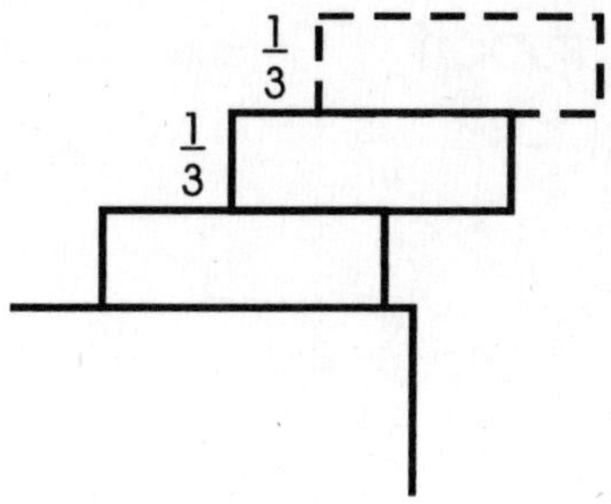

8. Two bricks of equal construction (length *l* and width *w*) are stacked at the edge of a table, as shown in the above diagram. If a third brick is stacked carefully on top, as shown by the dotted line,

(A) nothing will happen.
(B) the top brick will fall if the bricks are thick enough.
(C) the top brick and second brick will fall.
(D) the top two bricks will fall.
(E) the bricks will tip to drop only the top brick.

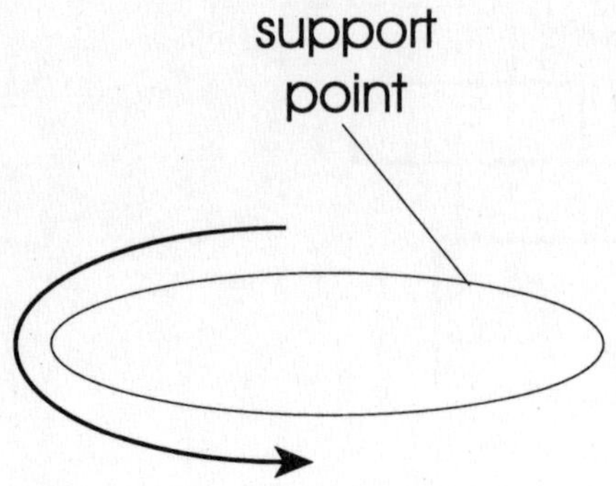

9. A rodeo rider (not pictured) swings a lariat overhead, as shown in the picture. The rope leading from the support point serves to drive the loop of rope below in a circle, as indicated by the arrow. The vector of centripetal force on a small piece of the loop of rope points

(A) along the direction of its motion.
(B) along the direction of the rope (toward the rider's hand).
(C) away from the direction of the rope.
(D) directly toward the rider.
(E) directly away from the rider.

10. Two bricks are placed on a table, one atop the other. The table has a positive coefficient of static friction μ_t between itself and the bricks, which are made of a frictional material and have a negative coefficient of kinetic friction μ_b between each other. When the top brick is nudged to the left a tiny bit,

(A) the top brick will slide to a stop almost instantly.
(B) the top brick will accelerate to the left.
(C) the top brick will accelerate left, and the bottom brick will slide right if

$$\frac{\mu_b}{\mu_t} < -2.$$

(D) the top brick will accelerate to the right.
(E) the top brick will accelerate to the left, and the bottom brick will slide right if

$$\frac{\mu_b}{\mu_t} > -\frac{1}{2}.$$

11. A truck (mass 1,000kg) and driver (mass 100 kg) can accelerate to 100 kph in 5 seconds. After picking up a 1,000 kg load of bricks, it takes 10 seconds. Assuming the engine exerts the same force, the driver can deduce that

(A) he was cheated and did not get 1,000 kg of bricks.
(B) he got too many bricks.
(C) he has exactly 1,000 kg of bricks.
(D) he initially drove uphill, but now he is driving downhill.
(E) none of the above are possible.

12. A baseball player crushes a home run deep over the far wall. The baseball bat

(A) exerts a greater force on the baseball than the baseball does on it.
(B) exerts less force on the baseball than the baseball does on it.
(C) exerts exactly the same force on the baseball as the baseball does on it.
(D) has no force exerted on it by the base ball, only by the player's arms.
(E) It is impossible to determine which exerted the greater force.

13. A box is slid at 5 m/s on a level surface with which it has a coefficient of kinetic friction $\mu_k = 1$. The box slides to a stop

(A) instantaneously.
(B) after sliding the same distance it was thrown.
(C) at a deceleration of *g*.
(D) after 3 seconds.
(E) after a distance of 3 m.

14. A monkey swings from a vine from one branch to another. The monkey's greatest total energy occurs

(A) at the top of the swing's arc.
(B) at the bottom of the swing's arc.
(C) halfway up the swing's arc.
(D) All of the above.
(E) None of the above.

15. A locomotive engine with mass 60 metric tons drives at 100 kph on straight, level, smooth tracks. In 1 km, the amount of work done is most closely

(A) 0 J.
(B) 60 kJ.
(C) 60 MJ.
(D) 60 GJ.
(E) 60 TJ.

16. A 900kg elevator accelerates at 1 m/s^2 upward. The total power required of the elevator motor is

(A) 900 W.
(B) 9,000 W.
(C) 9,900 W.
(D) constantly increasing.
(E) constantly decreasing.

17. A projectile is fired upward at 90 degrees. A similar projectile is fired at the same velocity at 60 degrees. The height attained by the second projectile is what fraction of the height of the first projectile?

(A) $\frac{1}{4}$

(B) $\frac{1}{3}$

(C) $\frac{1}{2}$

(D) $\frac{2}{3}$

(E) $\frac{3}{4}$

18. A projectile explodes at the top of its arc, splitting into two equal-mass pieces. The first piece stops from the explosion and falls straight to the ground. If the projectile had not exploded in flight it would have landed 1 km away. The second piece lands

(A) 1 km away.
(B) 1.5 km away.
(C) 2 km away.
(D) 2.5 km away.
(E) 3 km away.

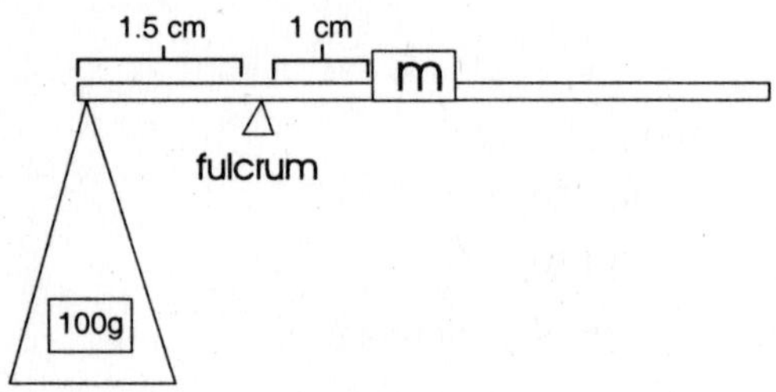

19. The material to be weighed on a balance hangs on the balance arm 1.5 cm from the fulcrum. The mass *m* rests in notches in the balance arm spaced 1 cm apart, denoting 100g increments on the other side of the fulcrum. The actual mass *m* is

(A) 50 g.
(B) 75 g.
(C) 100 g.
(D) 150 g.
(E) 200 g.

20. An 80 kg runner runs up a hill with a 30 degree slope at constant velocity. The minimum force that the runner's feet must apply along the road's surface to accomplish this is

(A) 400 kg.
(B) 800 kg.
(C) 1600 kg.
(D) 400 N.
(E) 800 N.

21. A small rubber ball bounces into a street in such a way that at the top of its last arc, it is motionless relative to anyone looking from the sidewalk. At that moment, a large and loaded semi tractor trailer moving at speed v collides with it elastically, sending the ball along the direction of motion of the truck. The velocity of the tiny ball after the collision is

(A) $3v$.
(B) $2v$.
(C) v.
(D) $\frac{v}{2}$.
(E) $\frac{v}{3}$.

22. A model rocket engine applies a total impulse of 10 N/s to a 300g model rocket in 0.1s. The final speed of the rocket is

(A) just over 0.03 m/s.
(B) just under 0.03 m/s.
(C) just over 30 m/s.
(D) just under 30 m/s.
(E) none of the above.

23. Two "fighting tops" of equal mass approach each other with low speeds and high angular momenta and collide elastically. If the tops are spinning at half the angular velocity after the collision, their final speeds will be

(A) zero.
(B) less than their initial speeds.
(C) the same as their initial speeds.
(D) greater than their initial speeds.
(E) exactly twice their initial speeds.

24. Two different people swing their legs through the same angle when they walk. One person is an adult with legs of length 2, ℓ. The other is a child with legs of length ℓ. Assuming their legs work like simple pendula and that the adult walks at speed v, the child walks at

(A) $\frac{v}{2}$.

(B) $\frac{v}{\sqrt{2}}$.

(C) $\frac{v}{3}$.

(D) $\frac{v}{\sqrt{3}}$.

(E) $\frac{v}{4}$.

25. A satellite is in circular orbit around the earth. A navigational error causes the satellite to enter a circular orbit where it collides with a similar satellite (also in circular orbit). Relative to the first satellite, the other satellite was traveling

(A) at the same speed as the first satellite.
(B) faster along its orbit.
(C) slower along its orbit.
(D) at the same speed as the first, opposite in direction.
(E) Collision is not possible.

26. A spacecraft in orbit performs an orbital maneuver so that it stops orbiting and supports itself by its rocket thrusters pointing at the Earth. A 160 lb. astronaut on the spacecraft steps on a scale and measures his weight to be 40 lbs. The radius of the Earth is R, so the spacecraft is

(A) R away from the Earth's surface.
(B) $2R$ from the surface.
(C) $3R$ from the surface.
(D) $4R$ from the surface.
(E) $5R$ from the surface.

27. A 100g marble is dropped into 1 kg of water so that it hits the surface traveling at its terminal velocity in the water. The marble travels 10 cm to the bottom, where it sticks. On its way through the water, the marble heats the water up by

(A) 1 μ K.
(B) 2.5 μ K.
(C) 10 μ K.
(D) 25 μ K.
(E) 100 μ K.

28. A car has 1m diameter rubber tires with a thermal expansion coefficient of 0.0001/K. After driving on the highway, the tires are 30 K warmer than when they started out. The new circumference of the tires is

(A) 1.0003 m.
(B) 1.003 m.
(C) π 1.0003 m.
(D) π 1.003 m.
(E) none of the above.

29. A driver realizes that her tires are over-pressurized to 42 PSI. She lets air out of the tires until the proper pressure for the car is reached. The air and tires are initially 30 degrees Celsius. In the nozzle, the gas expands to 2.5 times its original volume. Assuming an ideal gas, the temperature of the outgoing 14 PSI air is

(A) 100 K.
(B) 150 K.
(C) 200 K.
(D) 250 K.
(E) 300 K.

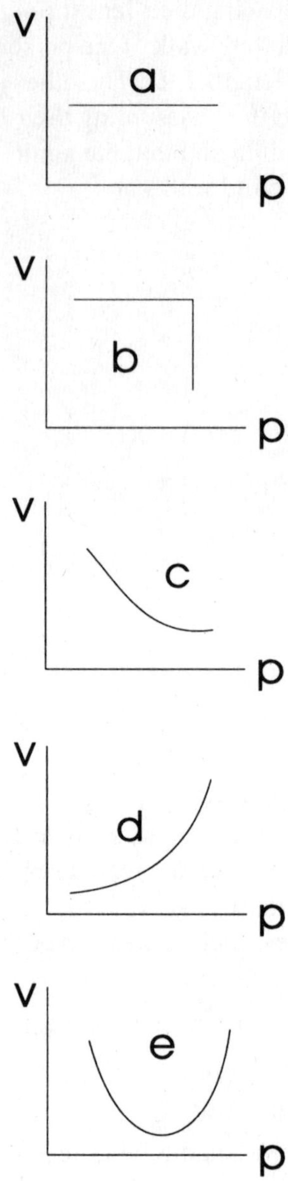

30. The expansion from question 29 is shown in which one of the P-V diagrams above?

(A) a
(B) b
(C) c
(D) d
(E) e

31. An imaginary heat engine cycle drawn on the P-V diagram consists of a rectangle with sides of lengths 20 N/m^2 and 0.1 m^3. The amount of work extracted from this cycle is

(A) 0.5 J.
(B) 1 J.
(C) 1.5 J.
(D) 2 J.
(E) 3 J.

32. A raisin at the bottom of a glass of soda water forms bubbles on its surface. The expanding bubbles lift the raisin to the surface and pop, letting the raisin fall back to the bottom. The diagram describing this process is

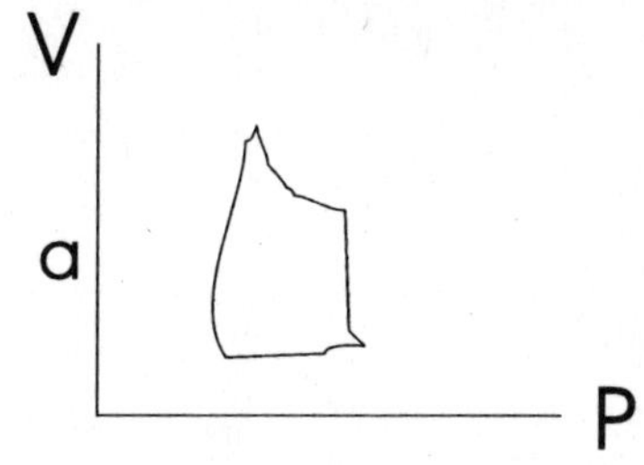

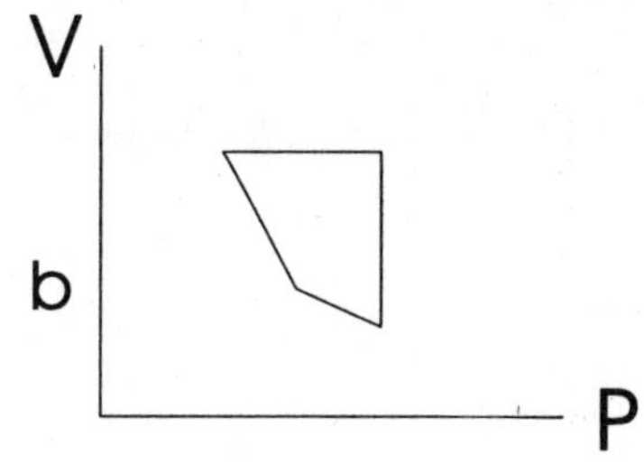

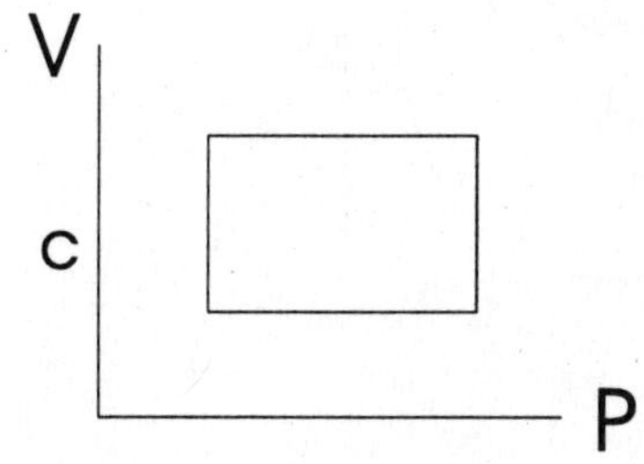

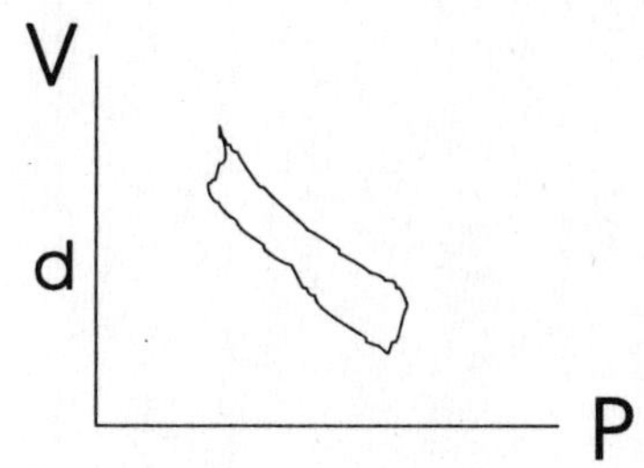

(A) a
(B) b
(C) c
(D) d
(E) None of the above.

33. A refrigerator light bulb with a broken-off switch delivers 40 W of power to the inside of a refrigerator. The refrigerator motor draws 1 A at 100 V and removes 80 W of heat from the refrigerator. The heat exhausted into the room is

(A) 120 W.
(B) 140 W.
(C) 180 W.
(D) 200 W.
(E) 220 W.

34. The magnitude of the net force on charge of q is

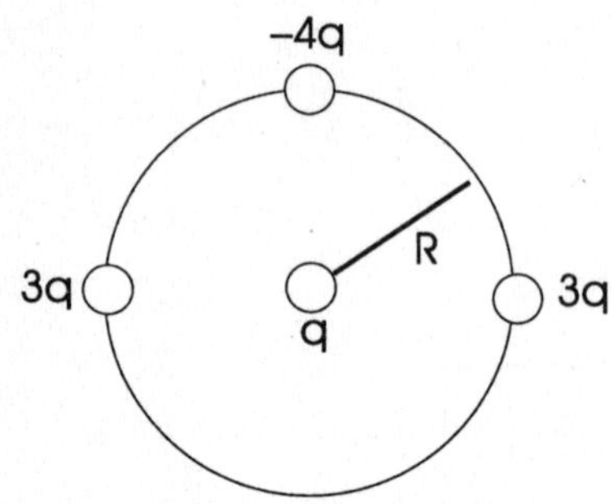

(A) $\frac{kq^2}{R^2}$

(B) $\frac{-kq^2}{R^2}$

(C) $\frac{-4kq^2}{R^2}$

(D) $\frac{4kq^2}{R^2}$

(E) $\frac{6kq^2}{R^2}$

35. What is the potential of a charge q (relative to its potential at an infinite distance perpendicular to the surface of the plane) 1m away from a truly infinite plane of charge?

(A) Infinite
(B) kq/m
(C) 10 kq/m
(D) 100 kq/m
(E) 1,000 kq/m

36. Two infinite planes of opposite charge densities (+ /- σ) are 1 m apart. The electric field 2 m from the negative plane is

(A) 0 N/C.

(B) $\frac{k\sigma}{4}$ N/C.

(C) $\frac{k\sigma}{2}$ N/C.

(D) $\frac{k\sigma}{3}$ N/C.

(E) $k\sigma$ N/C.

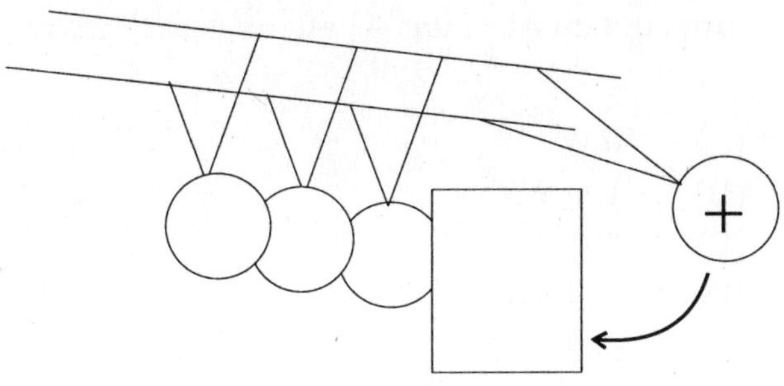

37. The first ball in a familiar collision toy (shown above) is shielded from the others by a thin sheet of plastic and is charged positively. All the balls are made of a conductive metal. When the ball swings down, the ball at the other end will swing up. It will be

(A) uncharged.
(B) charged negatively.
(C) charged positively.
(D) charged, but it will discharge as it swings up.
(E) impossible to tell how it will be charged.

38. The plates of a parallel plate capacitor are brought closer together. The voltage between the two plates

(A) increases.
(B) decreases.
(C) stays the same.
(D) stays the same, but the electric field increases.
(E) goes to zero.

39. Two plates of a parallel-plate capacitor are charged, then each is folded in half so that they still form a parallel-plate capacitor with half the area. The voltage and electric field between the two plates will

(A) increase, and the electric field will stay the same.
(B) decrease, and the electric field will stay the same.
(C) stay the same.
(D) increase, and the electric field increases.
(E) decrease, and the electric field decreases.

40. Voltage supplied to an old-fashioned lamp is converted into direct current. The voltage is 110 V and the bulb is a 100 W bulb. The resistance of the bulb filament is most nearly

(A) 10 Ω.
(B) 50 Ω.
(C) 60 Ω.
(D) 80 Ω.
(E) 121 Ω.

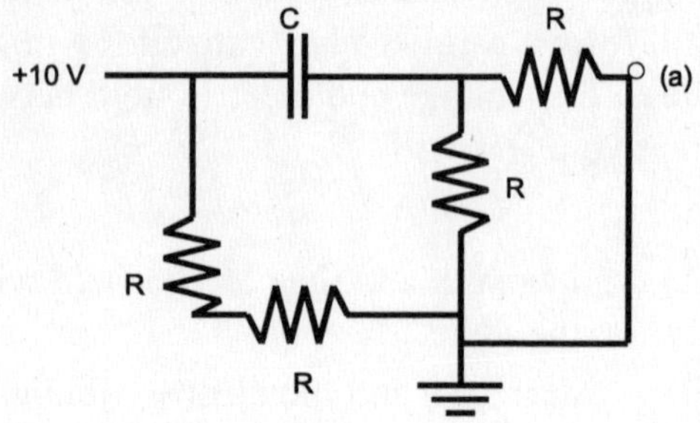

41. The circuit above is connected to a 10 V supply at one end and grounded at the other, as shown. The potential at point (a) is

(A) 0 V.
(B) 1 V.
(C) 2 V.
(D) 3 V.
(E) 4 V.

42. A capacitor and 10 Ω resistor in series are attached to a 100 V power supply. After a long time, the voltage across the capacitor is

(A) 0 V.
(B) 1 V.
(C) 10 V.
(D) 50 V.
(E) 100 V.

43. A capacitor and a 20 Ω resistor are connected to a 100 V power supply in parallel. The voltage across the capacitor after a long time is

(A) 0 V.
(B) 1 V.
(C) 10 V.
(D) 50 V.
(E) 100 V.

44. The difference in power used in the circuits from questions 42 and 43 after a long time is

(A) 0 W.
(B) 100 W.
(C) 500 W.
(D) 10,000 W.
(E) 50,000 W.

45. A car antenna is 1 m long. The frequency of radio waves it receives best is most closely

(A) 10 kHz.
(B) 300 kHz.
(C) 10 MHz.
(D) 300 MHz.
(E) 10 Ghz.

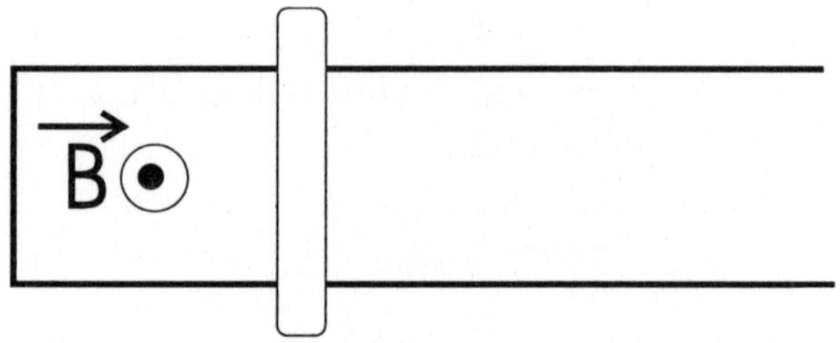

46. An electrically conductive metal bar slides frictionlessly on fixed, conductive rails, as shown in the diagram above. Assume the bar is actually a long distance from the left end of the loop. If the constant magnetic field shown is suddenly turned on (the initial field is zero), the bar will

(A) slide to the right and keep going.
(B) slide to the right and come to a stop.
(C) slide to the left and keep going.
(D) slide to the left and come to a stop.
(E) accelerate constantly to the right.

47. An electron is accelerated inside a TV tube and stops in a 700-atom-thick layer of phosphorus atoms. The atoms de-excite, emitting photons with roughly 0.4 eV each. If the accelerating voltage is one of the following, it is most likely

(A) 10 V.
(B) 270 V.
(C) 300 V.
(D) 1,000 V.
(E) 2,000 V.

48. The force on two parallel power lines conducting current in the same direction will

(A) pull them together.
(B) push them apart.
(C) do nothing.
(D) push one up and one down.
(E) Do none of the above.

49. An old television tube is positioned so that the magnetic field of the Earth passes through the screen perpendicular to the screen surface and pointing toward the back of the tube. This will

(A) shift the picture up.
(B) shift the picture down.
(C) rotate the picture.
(D) shift the picture left.
(E) shift the picture right.

50. A kitchen magnet falls off the refrigerator but stays near it as it accelerates toward the floor, 0.8 m away. Its magnetic field penetrates the metal refrigerator surface. Assuming g is exactly 10, by the time it reaches the floor it is traveling

(A) just under 4 m/s.
(B) exactly 4 m/s.
(C) just over 4 m/s.
(D) just under 2 m/s.
(E) just over 2 m/s.

51. An electromagnetic coil on a crane lifts a car up to be dropped into a car compactor for recycling. When the current in the coils is turned off, the car will

(A) just fall into the compactor.
(B) create a magnetic field that pushes it into the compactor.
(C) be cooler than when it was picked up.
(D) be warmer than when it was picked up.
(E) None of the above.

52. A spring has its ends connected with a wire. When a magnetic field is steadily turned up through the spring it will tend to

(A) oscillate faster and faster.
(B) oscillate slower and slower.
(C) compress slightly.
(D) stretch slightly.
(E) do nothing.

53. Two waves approach each other on a string, as shown below.

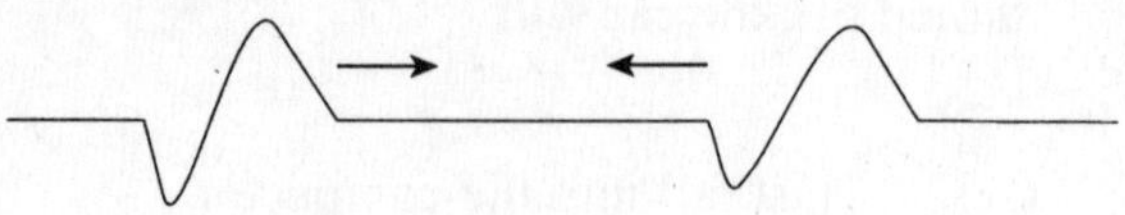

When they overlap completely, the wave looks like

(A)

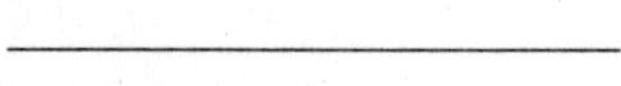

(B) 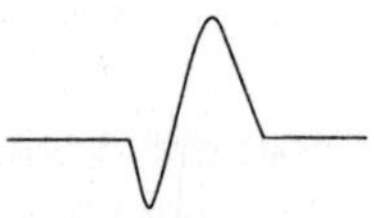

(C) 

(D) 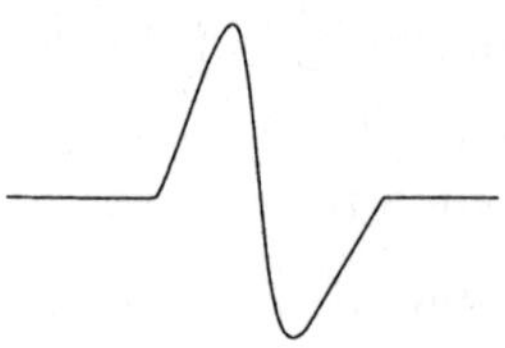

(E) 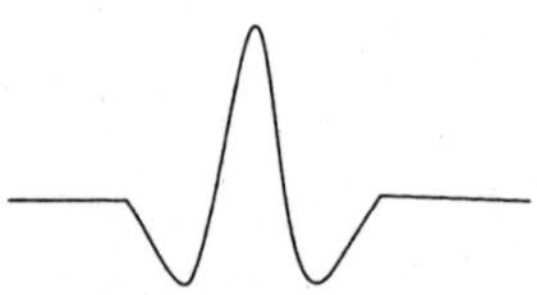

54. A siren on a police car is received at a higher frequency as the car approaches a listener at the side of the road. If the siren is turned around, as the car approaches the listener will hear

(A) a lower-pitched sound than before, but still higher than the driver hears.
(B) a lower-pitched sound than the driver hears.
(C) the same sound that the driver hears.
(D) the same sound the listener heard before the siren was turned.
(E) a higher-pitched sound than before.

55. Two perfectly flat, rectangular panes of glass are pressed together. The outer surfaces have been covered with antireflective coatings, and the surfaces in contact have been left untreated. Interference makes the glass appear

(A) dark.
(B) light.
(C) with horizontal fringes.
(D) with vertical fringes.
(E) with circular fringes.

56. A prism sits on a windowsill, creating a small rainbow on the floor. The glass in the prism creates a red band that is wider than orange, which is wider than yellow. Based on this observation, if we could see them,

(A) the infrared and ultraviolet bands would be narrower than the red band.
(B) the infrared and ultraviolet bands would be narrower than the violet band.
(C) the infrared bands would be wider than red, and the ultraviolet bands would be narrower than violet.
(D) the infrared would be narrower than red, and the ultraviolet would be wider than violet.
(E) they would both be the same size as green.

57. Two narrow slits, 0.01cm apart, are 1 m from a blank screen, as shown. A laser beam is incident upon the slits. The two first diffraction *minima* are how far apart?

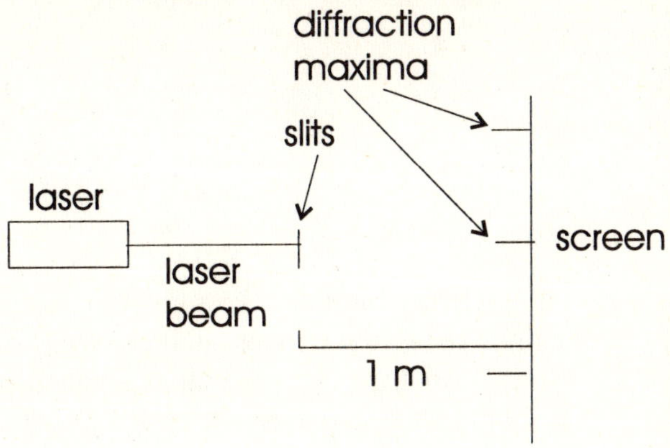

(A) 1 cm
(B) 1.5 cm
(C) 2 cm
(D) 2.5 cm
(E) It is impossible to tell.

58. A pool of water $\left(n = \frac{4}{3}\right)$ looks 1 m deep from the edge. In reality its depth is

(A) $\frac{9}{16}$ m.
(B) $\frac{3}{4}$ m.
(C) 1 m.
(D) $\frac{4}{3}$ m.
(E) $\frac{16}{9}$ m.

59. An object is projected through a simple convex lens. It is 1 m from the lens, and its image forms 1 m away. The focal length of the lens is

(A) 0.1 m.
(B) 0.5 m.
(C) 1 m.
(D) 2 m.
(E) 3 m.

60. A 4 cm tall object is placed 20 cm away from a concave mirror. The image of the object is 2 cm tall. The focal length of the mirror is

(A) 5 cm.
(B) $5\frac{2}{3}$ cm.
(C) 6 cm.
(D) $6\frac{2}{3}$ cm.
(E) 7 cm.

61. A layer of gold 1,000-atoms thick backscatters alpha particles (essentially point particles) incident upon it. Assume that backscattering only occurs if the particles make a direct hit upon the nucleus. If the nuclei are $\frac{1}{10,000}$ as wide as the atom, the percentage of backscattered alpha particles is approximately

(A) 0.00001%
(B) 0.001 %
(C) 1 %
(D) 10 %
(E) 100 %

62. A lightbulb filament in a copier develops a defect, causing it to emit twice as many red photons (800 nm) as it had previously emitted violet photons (400 nm). The photons normally reflect from copies and hit the metal ink roll, ejecting electrons and causing the ink to stick to the roll. The work function of the metal in the roll is 0.04 eV. If the violet photons had an energy of 0.06 eV, the copier will

(A) turn out blank sheets of paper.
(B) copy less darkly.
(C) not be affected.
(D) copy more darkly.
(E) turn out dark sheets of paper.

63. An electron binds to a hydrogen ion by jumping into an excited state emitting a photon of energy *E*. If it then jumps to the ground state, its orbit will be 1/4 its current orbit. The energy of the emitted photon would be

(A) $\frac{E}{3}$.
(B) $\frac{E}{2}$.
(C) *E*.
(D) 2 *E*.
(E) 3 *E*.

64. A tungsten lightbulb filament can emit a continuous spectrum, while a lamp of excited sodium atoms in a vapor emits only specific frequencies. This is because

(A) the tungsten can conduct electricity.
(B) tungsten atoms do not have specific energy levels.
(C) the tungsten atoms are larger.
(D) the tungsten atoms are smaller.
(E) the tungsten atoms are cooler.

65. In the extremely rare double-β decay process, atoms with proton number *Z* and neutron number *N* can emit two β-particles. After the transition, the particles contain

	protons	neutrons
(A)	Z	N.
(B)	$Z–2$	$N–2$.
(C)	$Z+2$	$N+2$.
(D)	$Z–2$	$N+2$.
(E)	$Z+2$	$N–2$.

66. A high-energy photon strikes an atom, removing one of the inner-shell electrons to infinity. Another electron makes a transition to the empty state, but does not emit a photon. Instead it could

(A) undergo radioactive decay.
(B) emit an electron.
(C) absorb a photon.
(D) do nothing.
(E) reabsorb the initial electron.

67. A 1 g radioactive sample reduces its absolute activity by 50% in one year, then only 30% the next year. The reason could be that

(A) the nuclei decayed into another unstable isotope.
(B) more than one radioactive substance was present.
(C) the half-life changed.
(D) B or C.
(E) A or B.

68. A nucleus will become less massive when it undergoes

(A) α-decay.
(B) β-decay.
(C) γ-decay.
(D) α or β-decay.
(E) any radioactive decay.

69. When a uranium nucleus splits in half (fission) and releases two neutrons, the halves' masses are

(A) slightly less than the original nucleus.
(B) slightly more than the original nucleus.
(C) less than the original nucleus by two neutron masses.
(D) more than the original nucleus by two neutron masses.
(E) less than the original nucleus by more than two neutron masses.

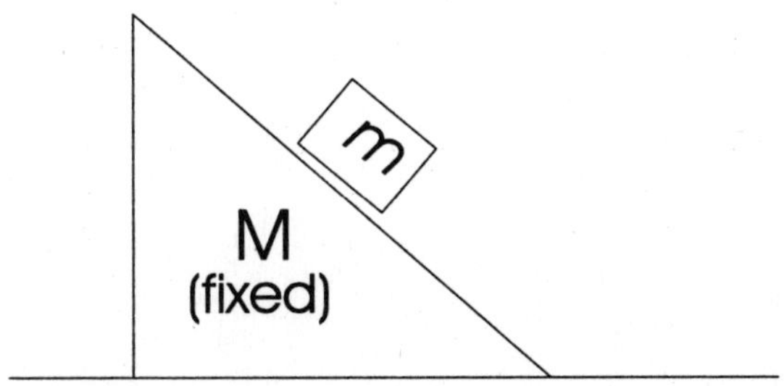

70. If the coefficient of friction between the block (mass *m*) and the inclined plane (mass *M*) is 0 (frictionless), then one may decrease the time it takes mass *m* to slide to the bottom of the plane by

(A) increasing *m*.
(B) increasing *M*.
(C) increasing the coefficient between the two blocks.
(D) decreasing the coefficient between the two blocks.
(E) none of the above.

SECTION II—FREE RESPONSE

Directions: Answer all six questions. Each question is designed to take approximately 15 minutes to complete. Note that subsections of the problem may not have equal weight. Show all work to obtain full credit, and avoid leaving important work on the green insert. (Assume g = 10m/s^2.)

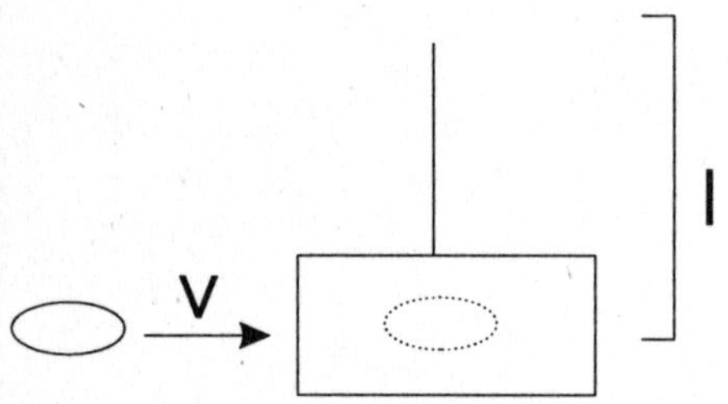

1. A ballistic pendulum (above) consists of a heavy cotton stopping-block (of mass *M*) suspended on a massless rod with its center of mass a distance *l* from the rod's pivot. A bullet of mass *m* is fired at a speed *v* horizontally into the exact center of the block, as shown. Express all answers algebraically in terms of *M*, *m*, *v*, *l*, and *g*.

   (A) Write an expression for the velocity, *V*, of the block immediately after the impact of the bullet.

   (B) Find the distance, *h*, the block will rise after the impact of the bullet.

   (C) Draw the vector of force, *F*, of the pivoting rod on its pivot point immediately after impact, and find its magnitude.

   (D) Assume that this gun is now fired horizontally at a distance of 1 m above level ground with no obstructions. Determine how far the bullet will travel.

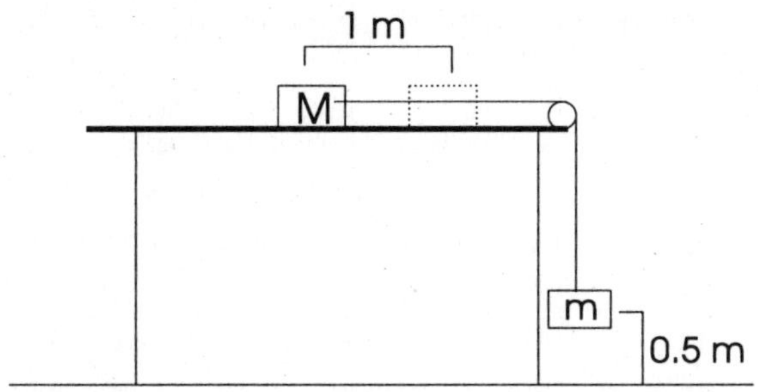

2. A mass m is attached via a massless string and pulley to another mass M, sitting on a table, as shown. The coefficient of kinetic friction between the block of mass M and the table is μ_k. The entire system is initially at rest. The block of mass m starts at a height of 0.5 m and descends to the floor, dragging mass M along the tabletop. Mass M slides to a stop 1 m from its original position. Express answers to the parts (A) and (B) in terms of M, m, and g.

   (A) Find an expression for the acceleration of mass m before it hits the floor.

   (B) Determine the coefficient of kinetic friction, μ_k, between mass M and the table.

   (C) Sketch a graph of the velocity of mass M versus time. Explain the graph in words and label any maxima or minima appropriately.

   (D) If $\mu_k = 0.2$, determine the mass M in terms of m.

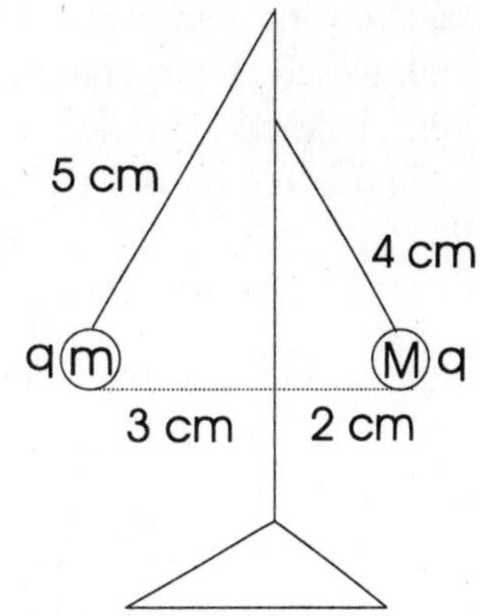

3. Two light pith balls hang together from strings of length 4 cm and 5 cm and from pivot points on a vertical axis in a classic demonstration on static electricity. When the balls are given a total charge $2q$, they will separate by 5 cm. The ball with the 4 cm string hangs closer (2 cm away) to the axis of separation than the other (3 cm away), as shown in the drawing above, and both hang at the same height. Assume that the balls receive equal charges q.

   (A) Draw all the forces on each ball in the space provided.

   (B) Determine the mass (m) of the ball with the 5 cm string in terms of the mass (M) of the near ball and gravity (g).

   (C) Now assume mass m is 0.1 g. Find the tension, T, in the string holding mass m.

   (D) Now mass on the long string is removed and the other ball (still charged) is left to swing in a magnetic field pointing along the vertical axis. Describe what will happen to the ball's motion.

4. A wind turbine drives a coil of wire with 200 turns and a 1 m^2 area about its diameter in the Earth's magnetic field, which is perpendicular to the axis of rotation. Assume the field strength of the Earth in this region is 1 Gauss (10^{-4} Tesla).

   (A) In order to light a lightbulb that requires a peak voltage of 120 V, find how fast the coil will have to spin.
   (B) If the coil is short-circuited and rotating at 100 rotations per second (its ends connected) and the resistance of the wire is 10 Ω, find how much heat it will generate.
   (C) Sketch the current in the coil as a function of time. Describe at maxima, minima, and zeroes which direction the coil is oriented in. The magnitude of any maxima or minima is unimportant.
   (D) If each coil was now separated and connected in its own separate ring, determine how the heat provided by the 20 separate rings would compare to the heat provided by the short-circuited coil in part (B).

5. A cylinder contains 3 *l* of an ideal gas at 1 atm and 300 K. The gas is first heated to 500 K at constant pressure. Then it is cooled at constant volume to 250 K. Then it is further cooled at constant pressure to 150 K. Finally, the gas is heated at constant volume to 300 K.

   (A) Sketch each step of this process in a *P*-*V* diagram, and give values for *P* and *V* at the end of each cycle. Label the lines 1–4 for the four steps in the problem.
   (B) Calculate the work done by this process.
   (C) Calculate the efficiency of this cycle, given an input heat energy of 600 J.
   (D) A small amount of water vapor is now introduced to the gas, and it condenses during the third step and evaporates again during the fourth step in the cycle. Describe how this effects the efficiency.

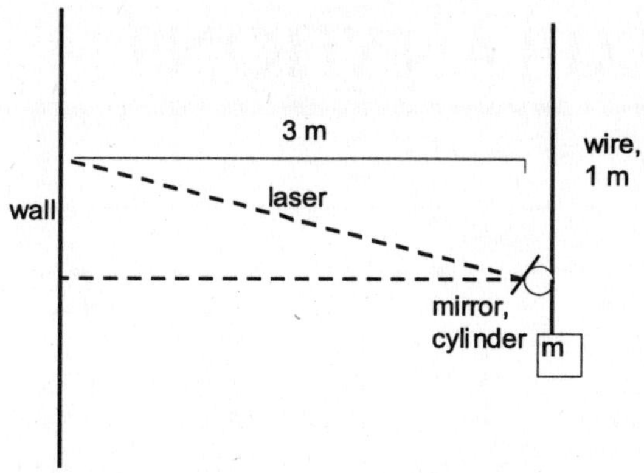

6. A wire is suspended from the ceiling and wrapped around a cylinder of 2 cm diameter that is free to rotate about its axis, as shown above. A small mass *m* provides some nominal tension. The length of wire between the ceiling and cylinder is 180 cm. Attached to the cylinder is a mirror that rotates with the cylinder if the wire stretches. A horizontal laser beam reflects from the mirror and strikes the wall, 3 m away, 40 cm above the beam height. The wire is now heated with an air gun by 40 K. The spot on the wall moves up as the wire expands by 5 cm.

   (A) From the movement of the laser spot, determine the coefficient of thermal expansion for the wire. You should keep angles in radians for this problem.

   (B) The expansion in part A means that some of the original wire is now wrapped around the cylinder, but the fixed end points are the same distance apart. Assuming the wire once had a mass density of 2.0 g/m, find the new mass density, and determine the change in the frequency of the first fundamental mode of the wire if the mass *m* suspended below is 100 g. If you cannot find the length of wire wrapped around the cylinder in part A, assume the amount of wire wrapped up is 0.01 cm and continue with this part.

Physics B Practice Test 1

ANSWERS AND EXPLANATIONS

SECTION I—MULTIPLE CHOICE

QUICK-SCORE ANSWERS									
1.C	8.A	15.A	22.D	29.D	36.A	43.E	50.A	57.E	64.A
2.A	9.D	16.D	23.D	30.C	37.C	44.C	51.D	58.D	65.E
3.D	10.C	17.E	24.B	31.D	38.B	45.D	52.C	59.B	66.B
4.E	11.B	18.B	25.A	32.E	39.D	46.D	53.B	60.D	67.E
5.D	12.C	19.D	26.A	33.E	40.E	47.C	54.D	61.B	68.E
6.A	13.C	20.D	27.D	34.D	41.A	48.A	55.A	62.A	69.E
7.E	14.E	21.B	28.D	35.A	42.E	49.C	56.C	63.E	70.E

1. **The correct answer is (C).** Tilting up causes the angular momentum vector of the propeller (which is pointing at the pilot) to have a downward component and a smaller component pointing toward the pilot. The plane will conserve angular momentum by tilting counterclockwise and turning left, creating new angular momentum vectors corresponding to motion of the main body of the plane opposite to the changes in the propeller's angular momentum vector.

2. **The correct answer is (A).** The formula $\mu = \tan(\theta)$ for a block sliding at constant velocity (not accelerating) is simply derived by Newton's second law: $\Sigma F = ma \Rightarrow \mu mg\cos(\theta) - mg\sin(\theta) = 0 \Rightarrow \mu = \tan(\theta) = \frac{3}{4}$.

3. **The correct answer is (D).** The plane must accelerate to catch the truck, then decelerate to match its velocity. The positions will then both intersect *and* have the same slope.

4. **The correct answer is (E).** Neither graph describes the motion. Changing down to positive makes the velocity correct but the position graph is backwards.

5. **The correct answer is (D).** The time of fall is $\sqrt{\frac{1}{4}} = \frac{1}{2}s$. The 2 m were covered in $\frac{1}{2}s$, so the velocity was 4 m/s.

6. **The correct answer is (A).** The bullet and the soldier fall at the same rate, so it will hit the soldier unless he hits the ground before it arrives. If he hits the ground first, it will hit the dirt and never get there. That will be in a little over a second, so it better be a low branch!

7. **The correct answer is (E).** Its displacement vectors (wind and air speed) form the legs of a 3-4-5 right triangle. The only sufficient wind speed is choice (E).

8. **The correct answer is (A).** Nothing will happen. The center of mass of the top two bricks is directly over the edge of the bottom brick.

9. **The correct answer is (D).** Centripetal force is the force that keeps the piece of rope moving in a circle and always points to the center of circular motion.

10. **The correct answer is (C).** A negative coefficient of friction would cause the creation of a force in the direction of motion, causing propulsion. The bottom brick would slide if the force provided by the friction can overcome its static friction with the table. Its normal force on the table is the weight of both bricks, so the negative coefficient of friction must be twice as large as its static coefficient of friction to allow this to happen.

11. **The correct answer is (B).** The increase in mass is the only explanation for the slow second acceleration. The new mass must be 1,100 kg to double the mass of the initial system (truck mass + driver mass), and if choice (D) were correct they would also be accelerating to 100 kph in less than 10 seconds on the second acceleration test.

12. **The correct answer is (C).** This is just an example of Newton's third law.

13. **The correct answer is (C).** The normal force is $\mu_k\,(mg) = mg$. Since $F = ma$, $mg = ma$, so $a = g$. The time to stop is 0.5 second, so the distance is 1.25 m.

14. **The correct answer is (D).** Total energy is conserved, choices (A), (B), and (C), and any other point in the swing has the same energy.

15. **The correct answer is (A).** The engine could be off, so the engine does no work. The speed and distance are superfluous because it's not going up hill or working against friction on a "smooth" track.

16. **The correct answer is (D).** The elevator may exert a constant force of 9,900 *N* on the cable, but the distance it travels per second is increasing with time, so the power increases. Each second it gets faster and has to do more and more work against gravity and its own acceleration.

17. **The correct answer is (E).** The first projectile has no kinetic energy at the top of its arc. The second projectile has half its initial velocity vector in the horizontal direction, so it has $\frac{1}{4}$ of its initial kinetic energy at the top of the arc. The rest ($\frac{3}{4}$ of the initial energy) has been converted into potential energy (mgh), and so it has attained a height of 3/4 that of the first projectile.

18. **The correct answer is (B).** The momentum of the second piece is equal to the initial momentum, but it has half the mass and therefore twice the velocity. The initial trip to the top of the arc is 0.5 km and takes the same time as the fall back to ground. In that time, the second piece will cover twice the distance (1 km) that it had originally, landing 1.5 km away.

19. **The correct answer is (D).** Static equilibrium is established if the torques balance, so for every 100 g added to the scale, the torque is 1.5 cm (100 g) 10 m/s^2 = 1,500 N-cm. Then mass m must move 1 cm away, adding a torque of 1(m) 10 m/s^2 = 1,500 N-cm. $m = 150$ g satisfies this equation and provides the necessary torque in the opposite direction.

20. **The correct answer is (D).** The runner's feet must supply a force, so all the answers with kg are out. To maintain constant velocity, the force provided by the feet along the road's surface must equal the force of gravity on the runner along the plane of the hill, or

$$m(g)\sin(30) = \frac{mg}{2} = 400N.$$

21. **The correct answer is (B).** The situation is like bouncing the ball off the truck at speed v. It would approach and recede from the truck at speed v because of the elastic collision. Since it would be receding from the truck at v and the truck would still essentially be moving at speed v, its total velocity would be 2 v.

22. **The correct answer is (D).** If you selected choices (A) or (B) you made the mistake of leaving the mass in grams and not converting to kilograms. Because the rocket will have moved up against gravity, it will be just under 30 m/s. (0.1 seconds is too short for the rocket to slow down significantly.)

23. **The correct answer is (D).** Conservation of energy in an elastic collision requires the angular kinetic energy to go somewhere, and that means into translation. The exact final speed depends on the exact angular momentum and initial speeds.

24. **The correct answer is (B).** The child will walk with strides that happen $\sqrt{2}$ times as often, but with half the stride length. $\frac{(\sqrt{2})}{2} = \frac{1}{\sqrt{2}}$, so the answer is choice (B).

25. **The correct answer is (A).** They could collide at an angle due to the orbit shift of the first satellite, but to be in circular orbit at the same height above the Earth they must be moving at the same speed in their orbits.

26. **The correct answer is (A).** By Newton's law of gravitation, the spacecraft is $2R$ away from the Earth's center, which puts it at R away from the surface.

27. **The correct answer is (D).** The work done by gravity is 0.1 kg (10 m/s^2) 0.1 m = 0.1 J. At 4.186 J/calorie, it deposits approximately $\frac{1}{40}$ of a calorie in the water, enough to heat the water

$\frac{1}{40{,}000}$K, or 25×10^{-6}K = 25mK.

28. **The correct answer is (D).** This was straightforward. The circumference of a circle of radius d is π (d), so the new circumference is just π (1.003) meters, after the thermal expansion. Not much of a change!

29. **The correct answer is (D).** The air is 300 K at the start. $pV = nRT$ after the expansion reads $p\left(\frac{1}{3}\right)V\left(\frac{5}{2}\right) = nRT_{new}$. T_{new} must then be equal to

$300\left(\frac{5}{6}\right) = 250\text{K}$.

30. **The correct answer is (C).** This shows an expansion (higher volume) and a drop in pressure. The drawing with a sudden kink showing the same would be a strange two-step process and not a single expansion.

31. **The correct answer is (D).** The work extracted by a heat engine is the area enclosed by its cycle diagram on the P-V plane. The area of the rectangle is 20 N/m^2 (0.1 m^3), or 2 J. Attention to units helps with this type of problem; the answer must be in N·m (or Joules), so the quantities must be multiplied in some way.

32. **The correct answer is (E).** This is not a closed process or cycle—the gas escapes. All the diagrams are closed cycles.

33. **The correct answer is (E).** Just add them all up. All that energy has to go somewhere, and it ends up in the room as exhaust.

34. **The correct answer is (D).** The forces from the left and right charges on the center charge cancel, so only the force from the top charge matters. Its magnitude is $\frac{4kq^2}{R^2}$, regardless of the direction.

35. **The correct answer is (A).** A truly infinite plane of charge has a constant electric field, so an infinite amount of work would have to be done to bring any charge to any finite distance (1 m, for example) from its surface.

36. **The correct answer is (A).** The fields will exactly cancel, and the field outside the gap will be 0.

37. **The correct answer is (C).** The approach of the first ball will draw negative charges near it, leaving the end ball charged positively. Its charge will not change as it swings up.

38. **The correct answer is (B).** The voltage decreases. The electric field from the plates of charge does not change, but the distance between the plates decreases. Much like a ball falling toward the Earth, the potential (mgh) decreases. In this case the electrical potential decreases when the attracting objects (oppositely charged plates) approach one another.

39. **The correct answer is (D).** The charge density increases (same charge in half the area), doubling the field. The distance stays the same, so the voltage between the two plates doubles.

40. **The correct answer is (E).**

$\frac{V^2}{R} = P.$ So $\frac{110^2 v}{R} = 100w$ so that R= 121 Ω

41. **The correct answer is (A).** Point (a) is connected to ground by a wire, so it is at 0 V.

42. **The correct answer is (E).** The capacitor will fully charge and current will stop flowing. That means no potential drop across the resistor, so the entire potential difference is across the capacitor.

43. **The correct answer is (E).** The capacitor will again charge, and current will flow through the resistor. But they are connected in parallel, so the entire 100 V potential change will be across each element.

44. **The correct answer is (C).** No current flows through the circuit in question 42, as explained in the previous answer, so no power is used. The power in the second circuit is given by $P = \frac{V^2}{R}$, or 500 W.

45. **The correct answer is (D).** The antenna will receive radio waves with wavelengths of approximately 1 m. The speed of light, *c*, is 300,000,000m/s. That means that it travels 1 meter $3*10^8$ times per second, giving it a frequency of 300 MHz.

46. **The correct answer is (D).** By Lenz's law, the direction of the field created by the new current must oppose the change in the magnetic flux through the opening. That means a clockwise current, creating a force on the bar and moving the bar to the left. As that loop closes, the flux will be decreasing, causing a counterclockwise current that will bring the bar to a stop.

47. **The correct answer is (C).** Conservation of energy leaves us an energy of 700*0.4 eV, or 280 eV given up by the electron and converted into photons. This is closest to 270 eV, but that means creating energy from nowhere. It is more likely 300 V, with the remaining 20 eV of energy lost heating the phosphor atoms.

48. **The correct answer is (A).** By applying two versions of the right-hand rule to first find the direction of the magnetic field of wire 1 on wire 2, and then finding the direction of the force on wire 2 by that field, we can see that it will be toward wire 1.

49. **The correct answer is (C).** The perpendicular magnetic field will move the picture depending on which way the electrons coming from the gun are directed (up, down, left, right). It won't shift the picture at all.

50. **The correct answer is (A).** The magnetic field will cause counteracting fields (Lenz's law) that lead to a slight magnetic braking. The speed in the absence of metal would have been exactly 4 m/s (from $mgh = \frac{1}{2}mv^2$).

51. **The correct answer is (D).** Eddy currents in the car in response to the changes in the magnetic field (picking it up and dropping it) will cause the car to heat up. The response field it creates would tend to stick it to the magnet longer, so choice (B) is wrong.

52. **The correct answer is (C).** The spring will get a current through it (Faraday's law), and since the wires conduct the current in parallel directions they will tend to attract, compressing the spring.

53. **The correct answer is (B).** Superimposing the waves tells us exactly how far up or down the individual pieces of string end up.

54. **The correct answer is (D).** The sound waves expand spherically, so it does not matter which way the emitter is pointed.

55. **The correct answer is (A).** The phase shift of 180 degrees at the front surface of the second pane of perfectly flat glass should cause uniform destructive interference across the entire surface.

56. **The correct answer is (C).** The logical conclusion would be that a band's width depends on its frequency—the lower the frequency, the wider the band. The infrared has the lowest frequency and the ultraviolet has the highest, leading to choice (C).

57. **The correct answer is (E).** You aren't given the wavelength, so you can't determine that information.

58. **The correct answer is (D).** The pool's apparent depth is given by our perception of the distance light traveled in the water. Because of the index of refraction of the water, the pool appears only $\frac{1}{n}$ as deep as it actually is.

59. **The correct answer is (B).** The lens equation in this case gives

$$\frac{1}{1}+\frac{1}{1}=2=\frac{1}{f}, \text{ thus, } f=0.5.$$

60. **The correct answer is (D).** $m=\frac{h_{\text{img}}}{h_{\text{obj}}}=\frac{d_{\text{img}}}{d_{\text{obj}}}$, so $d_{\text{img}}=10$ cm. Therefore, by the mirror equation, $\frac{1}{d_{\text{img}}}+\frac{1}{d_{\text{obj}}}=\frac{1}{f}=\frac{3}{20}$, which implies f is $6\frac{2}{3}$cm.

61. **The correct answer is (B).** The width of the nuclei gives them an area 10^{-8} that of the atom, but there are 1,000 of them in the layer. So the area of 1,000 nuclei is approximately 10^{-5} that of the area of an atom. They are small enough that we can assume little overlap. So an alpha particle has a $\frac{1}{10^5}$ chance of hitting one, or 0.001%.

62. **The correct answer is (A).** The photons of twice the wavelength have half the frequency and therefore half the energy. They don't have enough energy to overcome the work function of the metal and eject electrons, no matter how many of them there are.

63. **The correct answer is (E).** The radius of an electron's orbit in a Bohr orbit is proportional to n^2, where n is the state number it is in. The energy of the orbit is inversely proportional to the radius. The first transition would be from infinite n to an energy level $\frac{1}{4}$ that of the ground state. The second transition would give the other $\frac{3}{4}$ of the energy of a single transition to the ground state from infinity, or 3 times the first transition.

64. **The correct answer is (A).** Because it contains a few valence electrons that are free to move throughout the metal, the solid filament can emit a continuous spectrum. The individual atoms still have specific energy levels—all atoms do.

65. **The correct answer is (E).** β-decay makes a neutron into a proton. Double β-decay would do that twice, reducing N and increasing Z.

66. **The correct answer is (B).** The nuclear process of a radioactive decay is irrelevant, and the others cannot explain where the energy went. The atom can emit an electron (called an "Auger electron") from a higher shell, and its kinetic energy can carry away the excess energy.

67. **The correct answer is (E).** Either way, a second isotope with a longer half-life is necessary to explain the reduction in decay time. The isotope could have been present initially or been created by the decays of the initial sample.

68. **The correct answer is (E).** Any decay will lead to a state of lower energy and, therefore, lower mass. The ejection of a massive particle, such as an α, is not necessary.

69. **The correct answer is (E).** This is a highly energetic process, and the kinetic energy of the resultant particles reduces the final mass since matter was converted into energy in the process.

70. **The correct answer is (E).** Choice (D) is impossible, choice (C) will slow the block down, and the other choices will have no effect.

SECTION II—FREE RESPONSE

1. **(A)** Conservation of momentum leads to $(M+m)V = mv$, so

$$V = \frac{mv}{(M+m)}.$$

(B) Conservation of momentum gives us the velocity of the block after impact. During the impact, energy is lost, but after the impact it is all converted into gravitational potential energy. Therefore,

$(M+m)gh = \frac{1}{2}(M+m)V^2$, or $h = \frac{V^2}{2g}$. Using V from

part (A) of this question will yield $h = \frac{m^2v^2}{2g(M+m)^2}$.

(C) Immediately after impact, the rod supports the block and bullet. The resulting force on the pivot rod will obviously be straight down and have magnitude (M + m)g. The block-bullet system is now in circular motion, so it also must pull up with a force

$\frac{(M+m)V^2}{l}$ and, therefore, pulls on its pivot point with a force of this magnitude in the same direction, straight down. The forces

simply add up to give $|F| = (m+M)g + \frac{m^2v^2}{\ell(M+m)}$.

(D) The bullet will fall 1 m in $\sqrt{\left(\frac{1}{5}\right)}$ seconds, and is traveling at velocity v.

So it will travel a distance of $\frac{v}{\sqrt{(5)}}$ meters before it hits the ground.

2. **(A)** By Newton's laws, the sum of forces on the system must equal the mass of the system times its acceleration. Therefore, $mg - Mg\mu_k = (M+m)a$.

Solving this, we arrive at $a = \frac{g(m - M\mu_k)}{(m+M)}$. The units are correct, the units of mass cancel, and we are left with an acceleration (g).

(B) After a distance of 0.5 m, mass M is traveling at a velocity $v = \sqrt{(2a(0.5))}$, or $\sqrt{(a)}$. After 1 m, it slides to a halt at an acceleration of $g\mu_k$. Solving with our expression from part (A) of this question, $\mu_k = \frac{(m - M\mu_k)}{(M + m)}$. This requires some algebraic reduction, so we collect terms with μ_k on the left and get

$\mu_k\left[\frac{1+2M}{(m+M)}\right] = \frac{m}{(M+m)}$. Dividing, we arrive at

$$\mu_k = \frac{m}{\left(\frac{m+2M}{m+M}\right)(M+m)}, \text{ or } \mu_k = \frac{m}{(2M+m)}.$$

(C)

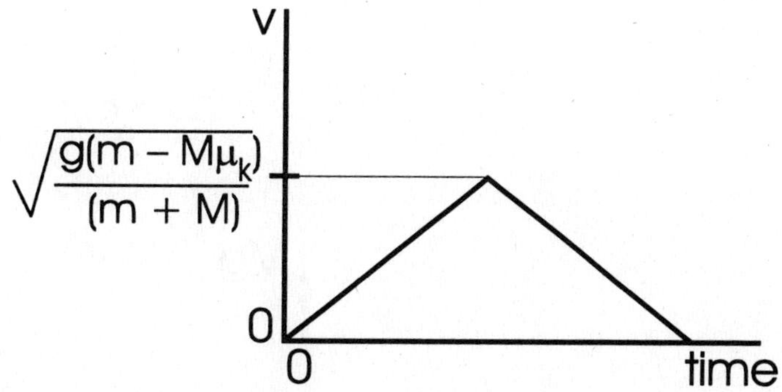

(D) Simplification in part (D) helps tremendously here, since the answer simplifies like so: $0.2(2M + m) = m$. $2M = 5m - m = 4m$, so $M = 2m$.

3. **(A)**

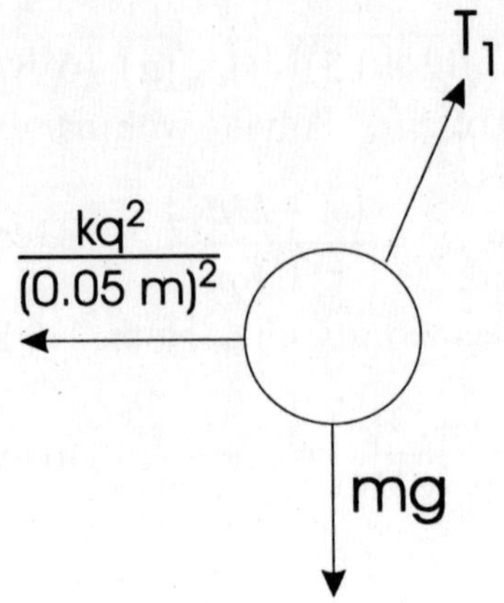

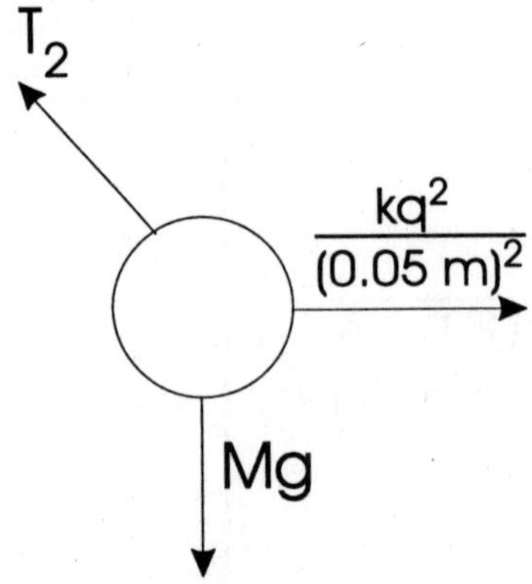

(B) Balancing the vertical and horizontal forces on m will yield: $T_1 \cos\phi = mg$ and $T_1 \sin\phi = \dfrac{kq^2}{(0.05m)^2}$, where ϕ is the angle of the string with the vertical axis. Eliminating T_1 will yield, $mg \tan\phi = \dfrac{kq^2}{(0.05m)^2}$. Doing the similar thing for M will yield $T_2 \cos\theta = Mg$ and $T_2 \sin\theta = \dfrac{kq^2}{(0.05m)^2}$, where θ is the angle of the string with the vertical axis. Eliminating T_2 will yield $Mg \tan\theta = \dfrac{kq^2}{(0.05m)^2}$. Since the forces of electrostatic repulsion are the same, $mg \tan\phi = \text{Mg} \tan\theta$, so after using $\tan\phi = \dfrac{3}{4}$ and q = 30°, we obtain $M = 1.3m$.

(C) $T\cos\phi = mg$, so that the sum of vertical forces is zero (the balls are stationary). Therefore, $T\left(\frac{4}{5}\right) = 0.001\ N$, and the magnitude of T is 0.00125 N.

(D) When the ball swings it will be pushed (since it is charged) to one side by its motion in the magnetic field. On the backswing it will be pushed to the other side. Since the period of a pendulum is relatively independent of the actual length of its swing, it will already be swinging back to that side. Eventually the pith ball will be slowly pushed into a circular orbit about the vertical axis in this fashion.

4. **(A)** $V = \frac{d\Phi}{dt}$. The amount of flux is sinusoidal, with maxima of $1m^2\ B$. The derivative is also sinusoidal (cosine function), with maxima ω B ($1m^2$) = 120 V. This is for a single coil. Adding the emf of all 200 together and solving for ω, we end up with 6,000 rad/s, or about 1 kHz (60,000 RPM). Clearly the Earth's magnetic field is a bad power source supply.

(B) From the previous subsection we can see that the coil will develop about 12 V across its ends. The heat output is given by $\frac{V^2}{R}$, or 14.4 Watts.

(C)

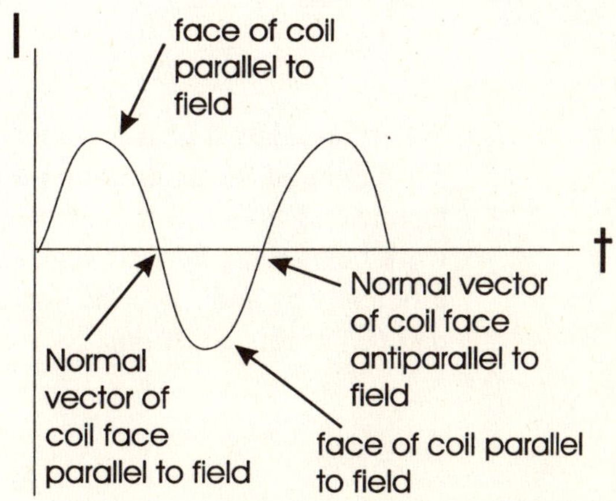

(D) Nothing will change. While the individual resistance of each coil will be less than that of the whole (since they are $\frac{1}{20}$ as long), the flux through each will also be $\frac{1}{20}$ of the flux through the entire coil, and so they will each carry the same current as before.

5. **(A)**

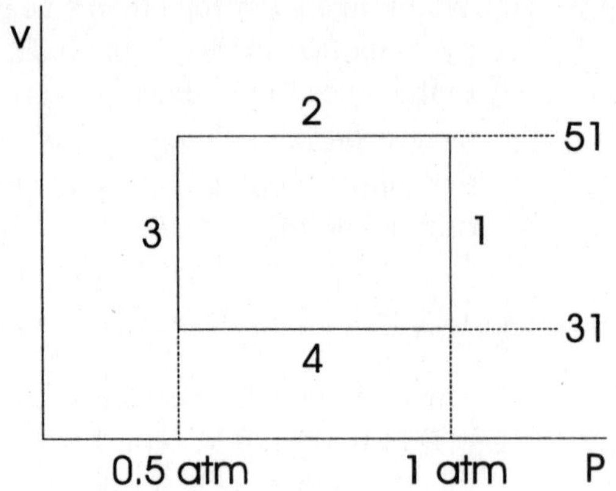

(B) The work done is the area under the curve, 1 ℓ-atm. 1 atmosphere is 100 kPa, and 1 l is 0.001 m^3, so the work done is 100 Joules.

(C) For this we need the heat going in, which is given as 600J. That means 500 J must be exhausted into the cold reservoir, for an efficiency of $\frac{1}{6}$.

$$\frac{(Q_{in} - Q_{out})}{Q_{in}} = \frac{W}{Q_{in}} = E \cdot$$

(D) Water has a latent heat of vaporization, so it will absorb extra heat during one of the heating-up part of the cycle (step 4) and will exhaust that heat again when it condenses in step 3. It will not really affect the work being done by the cycle, so its major effect would be to lower the efficiency by adding to the denominator of the algebraic expression in part (C) of this question.

6. **(A)** The beam reflection originally makes an angle given by the triangle in the diagram, with short sides of 3m and 0.4m. The inverse tangent gives an angle of 0.133 radians. After the heating, the angle is 0.149 radians. The angle change should be twice the angle change of the mirror (the beam reflecting from the mirror will change both its incident and reflected angles), or 0.008 radians. From the formula for the arc length of a circle, $s = r\theta$, we know that 0.008 cm of wire were wrapped around the cylinder. That means that $\frac{\Delta \ell}{\ell} = \frac{0.008}{180} = 4.44 * 10^{-5}$, and since $\Delta T = 40\ K$ the coefficient of expansion is $\frac{1.11 * 10^{-6}}{K}$.

   (B) This is somewhat easier. 0.008 cm of wire is wrapped around the cylinder, so the mass is less by 2g/m(0.00008 m). The mass density is less by $\frac{0.00016\text{g}}{1.8\text{m}} = 0.000088\text{g/m}$, so the new mass density is 1.999911g/m. The fundamental frequency of a wave on a stretched string of length L is $f = \left(\frac{1}{2}L\right)\sqrt{\left(\frac{T}{\mu}\right)}$. $T = 1N$ because of the 100 g mass, giving a fundamental frequency of $f = \frac{1}{\left(3.6\sqrt{\mu}\right)}$. The frequency difference in these strings is then (plugging in the mass densities and subtracting) 5.0 (10^{-4}) Hz. If you used 0.01 cm of wire, the new mass density (by the same method) would end up being 1.99988 g/m, and the answer would turn out to be 6.7 * 10^{-4} Hz.

ANSWER SHEET FOR PHYSICS B PRACTICE TEST II

Section I: Multiple Choice

1 ⊂A⊃ ⊂B⊃ ⊂C⊃ ⊂D⊃ ⊂E⊃
2 ⊂A⊃ ⊂B⊃ ⊂C⊃ ⊂D⊃ ⊂E⊃
3 ⊂A⊃ ⊂B⊃ ⊂C⊃ ⊂D⊃ ⊂E⊃
4 ⊂A⊃ ⊂B⊃ ⊂C⊃ ⊂D⊃ ⊂E⊃
5 ⊂A⊃ ⊂B⊃ ⊂C⊃ ⊂D⊃ ⊂E⊃
6 ⊂A⊃ ⊂B⊃ ⊂C⊃ ⊂D⊃ ⊂E⊃
7 ⊂A⊃ ⊂B⊃ ⊂C⊃ ⊂D⊃ ⊂E⊃
8 ⊂A⊃ ⊂B⊃ ⊂C⊃ ⊂D⊃ ⊂E⊃
9 ⊂A⊃ ⊂B⊃ ⊂C⊃ ⊂D⊃ ⊂E⊃
10 ⊂A⊃ ⊂B⊃ ⊂C⊃ ⊂D⊃ ⊂E⊃
11 ⊂A⊃ ⊂B⊃ ⊂C⊃ ⊂D⊃ ⊂E⊃
12 ⊂A⊃ ⊂B⊃ ⊂C⊃ ⊂D⊃ ⊂E⊃
13 ⊂A⊃ ⊂B⊃ ⊂C⊃ ⊂D⊃ ⊂E⊃
14 ⊂A⊃ ⊂B⊃ ⊂C⊃ ⊂D⊃ ⊂E⊃
15 ⊂A⊃ ⊂B⊃ ⊂C⊃ ⊂D⊃ ⊂E⊃
16 ⊂A⊃ ⊂B⊃ ⊂C⊃ ⊂D⊃ ⊂E⊃
17 ⊂A⊃ ⊂B⊃ ⊂C⊃ ⊂D⊃ ⊂E⊃
18 ⊂A⊃ ⊂B⊃ ⊂C⊃ ⊂D⊃ ⊂E⊃
19 ⊂A⊃ ⊂B⊃ ⊂C⊃ ⊂D⊃ ⊂E⊃
20 ⊂A⊃ ⊂B⊃ ⊂C⊃ ⊂D⊃ ⊂E⊃
21 ⊂A⊃ ⊂B⊃ ⊂C⊃ ⊂D⊃ ⊂E⊃
22 ⊂A⊃ ⊂B⊃ ⊂C⊃ ⊂D⊃ ⊂E⊃
23 ⊂A⊃ ⊂B⊃ ⊂C⊃ ⊂D⊃ ⊂E⊃
24 ⊂A⊃ ⊂B⊃ ⊂C⊃ ⊂D⊃ ⊂E⊃
25 ⊂A⊃ ⊂B⊃ ⊂C⊃ ⊂D⊃ ⊂E⊃
26 ⊂A⊃ ⊂B⊃ ⊂C⊃ ⊂D⊃ ⊂E⊃
27 ⊂A⊃ ⊂B⊃ ⊂C⊃ ⊂D⊃ ⊂E⊃
28 ⊂A⊃ ⊂B⊃ ⊂C⊃ ⊂D⊃ ⊂E⊃
29 ⊂A⊃ ⊂B⊃ ⊂C⊃ ⊂D⊃ ⊂E⊃
30 ⊂A⊃ ⊂B⊃ ⊂C⊃ ⊂D⊃ ⊂E⊃
31 ⊂A⊃ ⊂B⊃ ⊂C⊃ ⊂D⊃ ⊂E⊃
32 ⊂A⊃ ⊂B⊃ ⊂C⊃ ⊂D⊃ ⊂E⊃
33 ⊂A⊃ ⊂B⊃ ⊂C⊃ ⊂D⊃ ⊂E⊃
34 ⊂A⊃ ⊂B⊃ ⊂C⊃ ⊂D⊃ ⊂E⊃
35 ⊂A⊃ ⊂B⊃ ⊂C⊃ ⊂D⊃ ⊂E⊃
36 ⊂A⊃ ⊂B⊃ ⊂C⊃ ⊂D⊃ ⊂E⊃
37 ⊂A⊃ ⊂B⊃ ⊂C⊃ ⊂D⊃ ⊂E⊃
38 ⊂A⊃ ⊂B⊃ ⊂C⊃ ⊂D⊃ ⊂E⊃
39 ⊂A⊃ ⊂B⊃ ⊂C⊃ ⊂D⊃ ⊂E⊃
40 ⊂A⊃ ⊂B⊃ ⊂C⊃ ⊂D⊃ ⊂E⊃
41 ⊂A⊃ ⊂B⊃ ⊂C⊃ ⊂D⊃ ⊂E⊃
42 ⊂A⊃ ⊂B⊃ ⊂C⊃ ⊂D⊃ ⊂E⊃
43 ⊂A⊃ ⊂B⊃ ⊂C⊃ ⊂D⊃ ⊂E⊃
44 ⊂A⊃ ⊂B⊃ ⊂C⊃ ⊂D⊃ ⊂E⊃
45 ⊂A⊃ ⊂B⊃ ⊂C⊃ ⊂D⊃ ⊂E⊃
46 ⊂A⊃ ⊂B⊃ ⊂C⊃ ⊂D⊃ ⊂E⊃
47 ⊂A⊃ ⊂B⊃ ⊂C⊃ ⊂D⊃ ⊂E⊃
48 ⊂A⊃ ⊂B⊃ ⊂C⊃ ⊂D⊃ ⊂E⊃
49 ⊂A⊃ ⊂B⊃ ⊂C⊃ ⊂D⊃ ⊂E⊃
50 ⊂A⊃ ⊂B⊃ ⊂C⊃ ⊂D⊃ ⊂E⊃
51 ⊂A⊃ ⊂B⊃ ⊂C⊃ ⊂D⊃ ⊂E⊃
52 ⊂A⊃ ⊂B⊃ ⊂C⊃ ⊂D⊃ ⊂E⊃
53 ⊂A⊃ ⊂B⊃ ⊂C⊃ ⊂D⊃ ⊂E⊃
54 ⊂A⊃ ⊂B⊃ ⊂C⊃ ⊂D⊃ ⊂E⊃
55 ⊂A⊃ ⊂B⊃ ⊂C⊃ ⊂D⊃ ⊂E⊃
56 ⊂A⊃ ⊂B⊃ ⊂C⊃ ⊂D⊃ ⊂E⊃
57 ⊂A⊃ ⊂B⊃ ⊂C⊃ ⊂D⊃ ⊂E⊃
58 ⊂A⊃ ⊂B⊃ ⊂C⊃ ⊂D⊃ ⊂E⊃
59 ⊂A⊃ ⊂B⊃ ⊂C⊃ ⊂D⊃ ⊂E⊃
60 ⊂A⊃ ⊂B⊃ ⊂C⊃ ⊂D⊃ ⊂E⊃
61 ⊂A⊃ ⊂B⊃ ⊂C⊃ ⊂D⊃ ⊂E⊃
62 ⊂A⊃ ⊂B⊃ ⊂C⊃ ⊂D⊃ ⊂E⊃
63 ⊂A⊃ ⊂B⊃ ⊂C⊃ ⊂D⊃ ⊂E⊃
64 ⊂A⊃ ⊂B⊃ ⊂C⊃ ⊂D⊃ ⊂E⊃
65 ⊂A⊃ ⊂B⊃ ⊂C⊃ ⊂D⊃ ⊂E⊃
66 ⊂A⊃ ⊂B⊃ ⊂C⊃ ⊂D⊃ ⊂E⊃
67 ⊂A⊃ ⊂B⊃ ⊂C⊃ ⊂D⊃ ⊂E⊃
68 ⊂A⊃ ⊂B⊃ ⊂C⊃ ⊂D⊃ ⊂E⊃
69 ⊂A⊃ ⊂B⊃ ⊂C⊃ ⊂D⊃ ⊂E⊃
70 ⊂A⊃ ⊂B⊃ ⊂C⊃ ⊂D⊃ ⊂E⊃

Section I: Free Response

1 ⊂A⊃ ⊂B⊃ ⊂C⊃ ⊂D⊃ ⊂E⊃
2 ⊂A⊃ ⊂B⊃ ⊂C⊃ ⊂D⊃ ⊂E⊃
3 ⊂A⊃ ⊂B⊃ ⊂C⊃ ⊂D⊃ ⊂E⊃
4 ⊂A⊃ ⊂B⊃ ⊂C⊃ ⊂D⊃ ⊂E⊃
5 ⊂A⊃ ⊂B⊃ ⊂C⊃ ⊂D⊃ ⊂E⊃
6 ⊂A⊃ ⊂B⊃ ⊂C⊃ ⊂D⊃ ⊂E⊃

Physics Formulas

TABLE OF INFORMATION

Constants and Conversion Factors

1 unified atomic mass unit.	$1 \text{ u} = 1.66 \times 10^{-27}$ kg = 931 MeV / c^2
Proton mass	$m_p = 1.67 \times 10^{-27}$ kg
Neutron mass	$m_n = 1.67 \times 10^{-27}$ kg
Electron mass	$m_e = 9.11 \times 10^{-31}$ kg
Magnitude of electron charge	$e = 1.60 \times 10^{-19}$
Avogadro's number	$N_0 = 6.02 \times 10^{23}$ mol^{-1}
Universal gas constant	R = 8.31 J / (mol × K)
Boltzmann's constant	$k_B = 1.38 \times 10^{-23}$ J/K
Speed of light	$c = 3.00 \times 10^8$ m/s
Planck's constant	$h = 6.63 \times 10^{-34}$J × s = 4.14×10^{-15} eV × s
	Hc = 1.99×10^{-25}J × m = 1.24×10^3 eV × nm
Vacuum permittivity	$e_0 = 8.85 \times 10^{-12}$C2/N × m^2
Coulomb's law constant	$k = 1/4\pi e_0 = 9.0 \times 10^9$N × m^2/C^2
Vacuum permeability	$m_0 = 4p \times 10^{-7}$Wb/(A × m)
Magnetic constant	$k' = m_0/4\pi \times 10^{-7}$Wb/(A × m)
Universal gravitational constant	$G = 6.67 \times 10^{-11}$ m^3/kg × s^2
Acceleration due to gravity at the Earth's surface	g = 9.8 m/s^2
1 atmosphere pressure	1 atm = 1.0×10^5 N/m^2 = 1.0×10^5 Pa
1 electron volt	1 eV = 1.60×10^{-19}J
1 angstrom	1 Å = 1×10^{-10} m

Units

Name	Symbol
meter	m
kilogram	kg
second	s
ampere	A
kelvin	K
mole	mol
hertz	Hz
newton	N
pascal	Pa
joule	J
watt	W
coulomb	C
volt	V
ohm	W
henry	H
farad	F
weber	Wb
tesla	T
degree Celsius	°C
electron-volt	eV

Prefixes

Factor	Prefix	Symbol
10^9	giga	G
10^6	mega	M
10^3	kilo	k
10^{-2}	centi	c
10^{-3}	milli	m
10^{-6}	micro	m
10^{-9}	nano	n
10^{-12}	pico	p

Values of Trigonometric Functions For Common Angles

Angle	Sin	Cos	Tan
0°	0	1	0
30°	1 / 2	$\sqrt{3}/2$	$\sqrt{3}/3$
37°	3 / 5	4 / 5	3 / 4
45°	$\sqrt{2}/2$	$\sqrt{2}/2$	1
53°	4 / 5	3 / 5	4 / 3
60°	$\sqrt{3}/2$	1 / 2	$\sqrt{3}$
90°	1	0	∞

Newtonian Mechanics

a = acceleration	F = force
f = frequency	h = height
J = impulse	K = kinetic energy
k = spring constant	l = length
m = mass	N = normal force
P = power	p = momentum
r = radius or distance	s = displacement
T = period	t = time
U = potential energy	v = velocity or speed
W = work	x = position

Physics B

PRACTICE TEST 2

SECTION I—MULTIPLE CHOICE

Directions: Each question listed below has five possible choices. Select the best answer given the information in each problem, and mark the corresponding oval on the answer sheet. (You may assume $g = 10\ \text{m/s}^2$).

1. Two cars collide in an intersection and lock together. Which of the following is conserved?

   (A) kinetic energy
   (B) momentum
   (C) total energy
   (D) choices (A) and (B)
   (E) choices (B) and (C)

2. A 40kg object initially moving at 5 m/s slides to rest in 5 *m* on a horizontal surface. What is the coefficient of kinetic friction between the object and the surface?

   (A) 0.1
   (B) 0.25
   (C) 0.4
   (D) 0.5
   (E) 1.0

3. A mass is moving at a velocity *v* and is subject to a constant acceleration *a*, as shown in the figure below. Which vector could not possibly represent the velocity at a later time?

   (A) 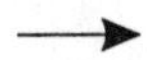
   (B) 
   (C) $v = 0$
   (D) 
   (E) 

4. A ball is dropped from a position of rest. Which of the plots shown below represent the speed as a function of displacement, *s*, from the initial position?

(A) 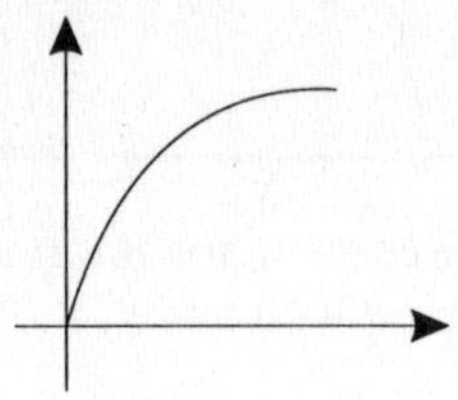

(B) 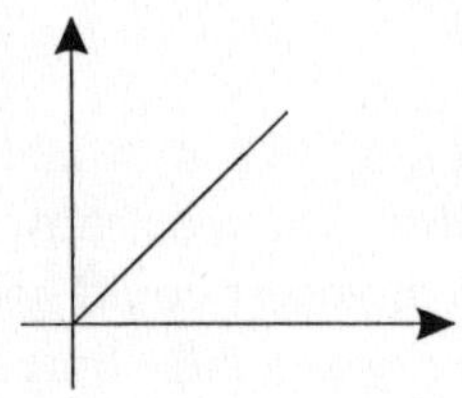

(C) 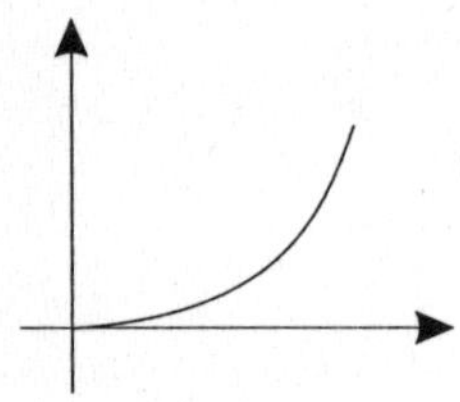

(D) 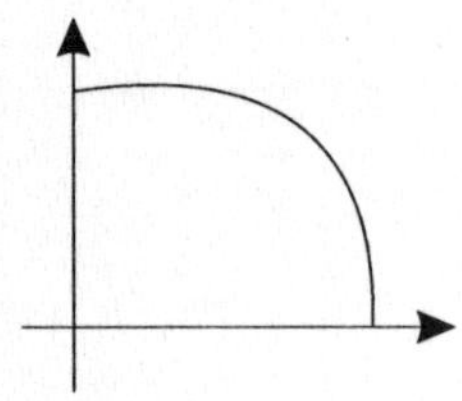

(E) 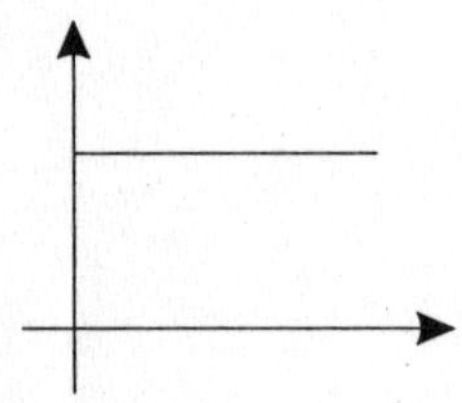

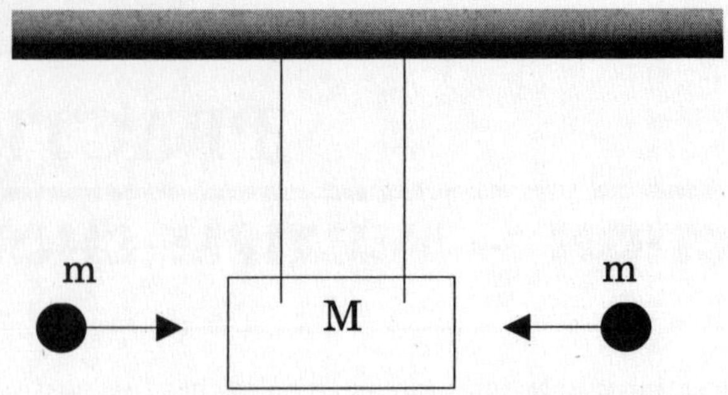

5. A gun manufacturer uses a ballistic pendulum to compare the muzzle speeds of the bullets fired from two guns. The bullets (mass *m*) are fired from opposite directions into a pendulum of mass *M,* as shown in the figure above. If the pendulum swings to a maximum height, *h,* above its initial position, what is the difference in speed of the two bullets?

(A) $\frac{M}{m}\sqrt{2gh}$

(B) $\frac{2m+M}{m}\sqrt{2gh}$

(C) $\frac{M}{m}2gh$

(D) $\frac{2m+M}{m}2gh$

(E) $\frac{m}{2m+M}\sqrt{2gh}$

6. A projectile is fired from a cannon horizontally from the edge of a 125 m high cliff at 200 m/s. How far from the bottom of the cliff does the projectile land?

(A) 125 m
(B) 200 m
(C) 400 m
(D) 1,000 m
(E) 1,250 m

7. Melissa is pushing horizontally on a box of width *w*, height *h*, and uniform mass distribution. At what angle from the vertical will the box just tip over?

(A) $\sin^{-1}\frac{w}{h}$

(B) $\sin^{-1}\frac{h}{w}$

(C) $\tan^{-1}\frac{w}{h}$

(D) $\tan^{-1}\frac{h}{w}$

(E) $\frac{w}{h}$

8. A 10 kg mass is acted upon by a 4 *N* force parallel to the *x*-axis and a 3 *N* force parallel to the *y*-axis. The *x*-*y* plane is horizontal. What is the magnitude of the acceleration of the mass?

(A) 0.3 m/s^2
(B) 0.5 m/s^2
(C) 0.7 m/s^2
(D) 3 m/s^2
(E) 5 m/s^2

9. Jack is standing on a 20 kg boat at rest in a lake. He jumps off the boat with a horizontal velocity of 3 m/s. If Jack's mass is 80 kg, what is the speed of the boat immediately after he jumps?

(A) 0.3 m/s
(B) 0.75 m/s
(C) 6 m/s
(D) 12 m/s
(E) 24 m/s

10. A ball is thrown straight up. When it reaches its maximum height,

(A) the acceleration is zero.
(B) the velocity is zero.
(C) the acceleration changes sign.
(D) choices (A) and (B)
(E) none of the above

11. A 40g ball bounces from a wall without losing mechanical energy. If its speed is 10 m/s just before making contact with the wall, and the force of the wall on the ball is equal to 16 *N* while ball and wall are in contact, how long were they in contact?

(A) 20 ms
(B) 25 ms
(C) 50 ms
(D) 100 ms
(E) 250 ms

12. Dave carries a 20kg box to the top of a 30m tall building. Only then does he realize that it actually belongs on the sixth floor, which is exactly halfway to the top of the building. After he delivers the box to the sixth floor, how much work has he done on the box?

(A) 600 J
(B) 900 J
(C) 3,000 J
(D) 6,000 J
(E) 9,000 J

13. Bill and Al are moving a piano. When it gets stuck on the front lawn, they push on it for 30 seconds, each with a force of 600N, without making it move. How much work have they done on the piano?

(A) 0 J
(B) 600 J
(C) 1,200 J
(D) 18,000 J
(E) 36,000 J

14. Abby is swinging a ball on a string in a circle at a constant speed. Which of the following quantities is constant?

(A) Acceleration
(B) Velocity
(C) Force
(D) Speed
(E) None of the above

15. Barbara is pushing horizontally a 15 kg carton across the floor. The coefficient of friction is 0.2. If the carton is accelerating at 0.3 m/s^2, how much force is Barbara exerting?

(A) 4.5 N
(B) 30 N
(C) 34.5 N
(D) 45 N
(E) Zero

16. David is lifting a 20 kg package by pulling straight up with 250 N. What is the acceleration of the package?

(A) 2.5 m/s^2
(B) 5 m/s^2
(C) 10 m/s^2
(D) 12.5 m/s^2
(E) 25 m/s^2

17. Elaine is riding in an elevator while standing on a bathroom scale. The scale reads 60 kg when the elevator is stopped. What does the scale read when the elevator is accelerating down at 1 m/s^2?

(A) 52 kg
(B) 54 kg
(C) 60 kg
(D) 66 kg
(E) 72 kg

18. A pendulum, whose period is one second on Earth, is taken to Mars, where the gravitational acceleration is 40% of the value on Earth. Approximately what is the period on Mars?

(A) 0.4 s
(B) 0.6 s
(C) 1 s
(D) 1.6 s
(E) 2.5 s

19. A 200 kg car is traveling at 50 m/s on a level, circular track with a radius of 100 m. What is the traction force required to keep the car on the track?

(A) 2,500 N
(B) 5,000 N
(C) 10,000 N
(D) 25,000 N
(E) 50,000 N

20. When a ball is released from a height of 100 cm, it only rises to 80 cm on the next bounce. Assuming the ball loses the same fraction of its own mechanical energy on each bounce, how high will it rise after the second bounce?

(A) 50 cm
(B) 60 cm
(C) 64 cm
(D) 68 cm
(E) 70 cm

21. A truck is pulling a 300 kg box across a flat, level surface at 2 m/s. The coefficient of friction between the box and the surface is 0.2. What power must the truck provide to pull the box?

(A) 200 W
(B) 300 W
(C) 600 W
(D) 1,200 W
(E) 3,000 W

22. A spring is compressed between two 5 kg masses at rest. When the spring is released, the masses travel apart at 2 m/s. What is the total momentum of the masses?

(A) 4 N–s
(B) 10 N–s
(C) 20 N–s
(D) 40 N–s
(E) None of the above

23. A 2 kg mass is attached to a vertical, uncompressed spring with a spring constant of 50 N/m. How far will the mass fall when released before bouncing up?

(A) 0.4 m
(B) 0.5 m
(C) 0.8 m
(D) 1 m
(E) 1.6 m

24. A mass on a spring oscillates with a period T. If the mass is doubled, the new period is

(A) $\frac{T}{2}$.
(B) $\frac{T}{\sqrt{2}}$.
(C) T.
(D) $T\sqrt{2}$.
(E) $2T$.

25. A penny is placed on a disk that is rotating at 2 revolutions per second. The coefficient of friction between the penny and the disk is 0.4. What is the maximum radius at which the penny will stay on the disk?

(A) 1 cm
(B) 2.5 cm
(C) 10 cm
(D) 25 cm
(E) 100 cm

26. A 3kg. mass is oscillating on a spring with an amplitude of 14 cm and a period of 2 sec. At what displacement from equilibrium is the kinetic energy equal to the potential energy?

(A) 0
(B) 5 cm
(C) 7 cm
(D) 10 cm
(E) 14 cm

27. An ice cube is floating in water in thermal equilibrium. Heat is gradually added to the water without disturbing thermal equilibrium. Which of the following occurs initially?

(A) The ice temperature rises.
(B) The water temperature rises.
(C) Some of the ice melts.
(D) Choices (A) and (B)
(E) Choices (A), (B), and (C)

28. Water is poured into an insulated container from a height of 60 m. How much does the temperature of the water increase? Assume the specific heat of water is $\frac{4\text{J}}{(g - °\text{C})}$.

(A) 0.03 °C
(B) 0.15 °C
(C) 0.3 °C
(D) 1.5 °C
(E) 3 °C

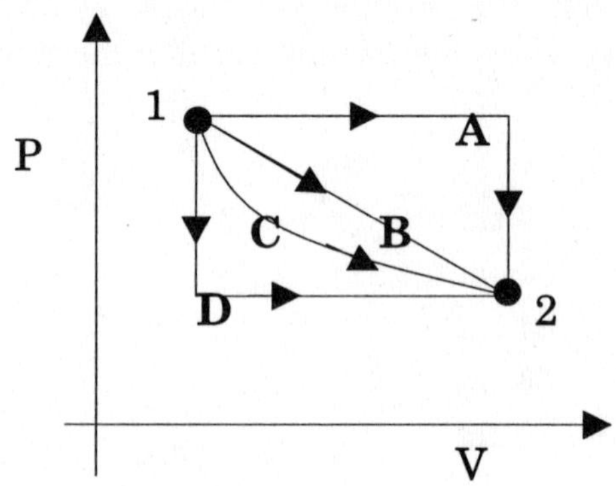

29. In the P–V diagram of a gas shown above, on which path from 1 to 2 is the most work done by the gas?

(A) A
(B) B
(C) C
(D) D
(E) All paths require the same work.

30. Consider a refrigerator that removes heat from its interior and exhausts an equal amount into the kitchen. This device is

(A) not allowed by the first law of thermodynamics.
(B) not allowed by the second law of the thermodynamics.
(C) not allowed by the third law of thermodynamics.
(D) allowed only if the total entropy increases.
(E) allowed if friction is neglected.

31. An ideal gas is heated in a closed, rigid container so its temperature doubles. The pressure

(A) decreases to half its initial value.
(B) remains constant.
(C) increases by $\sqrt{2}$.
(D) doubles.
(E) cannot be determined from the information given.

32. A heat engine withdraws heat from a reservoir at 500°C and exhausts it at 100°C. The temperature of the hot reservoir is increased to 600°C. Which of the following statements is true?

(A) The maximum efficiency will increase.
(B) The maximum efficiency will decrease.
(C) The exhaust temperature will increase.
(D) The exhaust temperature will decrease.
(E) The engine will run faster.

33. An ideal gas is compressed in a thermally isolated chamber. Which of the following occurs?

(A) The pressure increases.
(B) The temperature increases.
(C) The temperature remains constant.
(D) Choices (A) and (B)
(E) Choices (A) and (C)

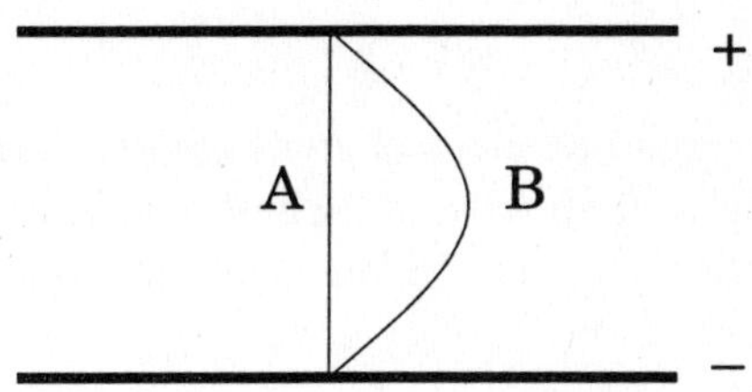

34. A positive charge is moved from the bottom (negative) plate to the top (positive) plate of the capacitor depicted above by either the path *A* or *B*. The work required to move the charge is

(A) positive and greater for *B* than for *A*.
(B) negative and greater for *B* than for *A*.
(C) positive and greater for *A* than for *B*.
(D) negative and greater for *A* than for *B*.
(E) positive and equal for *A* and *B*.

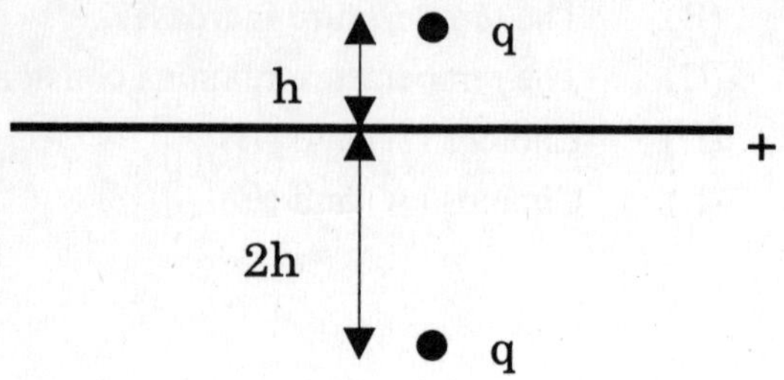

35. Two equal positive charges are positioned a distance *h* above and 2*h* below an infinite, positively charged plate, as depicted above. The forces on the charges are

(A) equal and in the same direction.
(B) equal and in opposite directions.
(C) twice as large for the upper charge and in the same direction.
(D) four times as large for the upper charge and in the same direction.
(E) four times as large for the upper charge and in opposite directions.

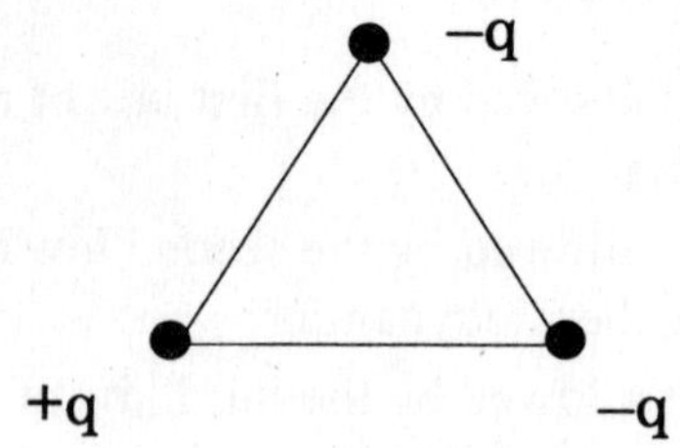

36. Three charges are placed at the vertices of an equilateral triangle, as shown in the figure above. The force on the upper charge is

(A) to the right.
(B) to the left.
(C) up.
(D) down.
(E) dependent on the sign of *q*.

37. An electric charge is placed near a neutral, conducting sphere. The force in the charge is

(A) zero.
(B) repulsive.
(C) attractive.
(D) attractive if the charge is positive, repulsive if the charge is negative.
(E) attractive if the charge is negative, repulsive if the charge is positive.

38. An oil drop of mass 10^{-12} grams is falling between two neutral, conducting plates spaced at 10 cm. The plates are suddenly charged to produce a potential difference of 1,000 V between the plates; the upper plate is negatively charged. If a drop has a net charge of $+10^{-18}$ coulombs, what is the acceleration of the drop when the plates are charged?

(A) $+10\ \text{m/s}^2$
(B) $+5\ \text{m/s}^2$
(C) 0
(D) $-5\ \text{m/s}^2$
(E) $-10\ \text{m/s}^2$

39. The distance between two charges, q_1 and q_2, is changed from r_1 to r_2. If k is the Coulomb Law constant, the energy required to move the charges is

(A) $k\frac{q_1}{r_1} - k\frac{q_2}{r_2}$.

(B) $k\frac{q_2}{r_2} - k\frac{q_1}{r_1}$.

(C) $k\frac{q_1q_2}{r_1^2} - k\frac{q_1q_2}{r_2^2}$.

(D) $k\frac{q_1q_2}{r_2} - k\frac{q_1q_2}{r_1}$.

(E) $k\frac{q_1q_2}{r_1^2} - k\frac{q_1q_2}{r_2}$.

<u>Questions 40 and 41</u> refer to the following figure.

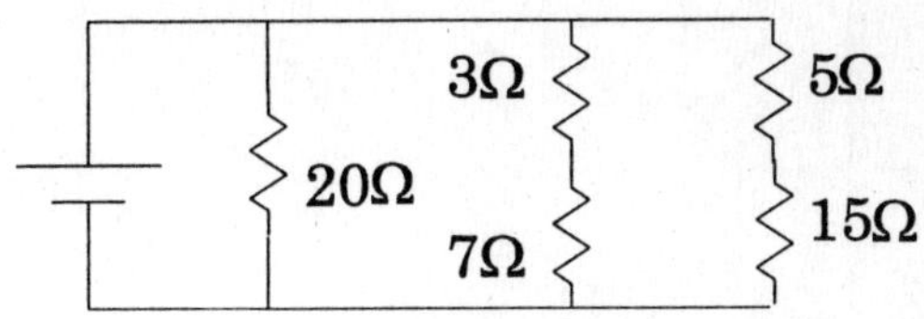

40. Which resistor dissipates the most power?

(A) 20 Ω
(B) 3 Ω
(C) 7 Ω
(D) 10 Ω
(E) 15 Ω

41. If the battery voltage is 10 V, how much current does it provide to the circuit?

(A) 0.2 A
(B) 0.5 A
(C) 1 A
(D) 2 A
(E) 5 A

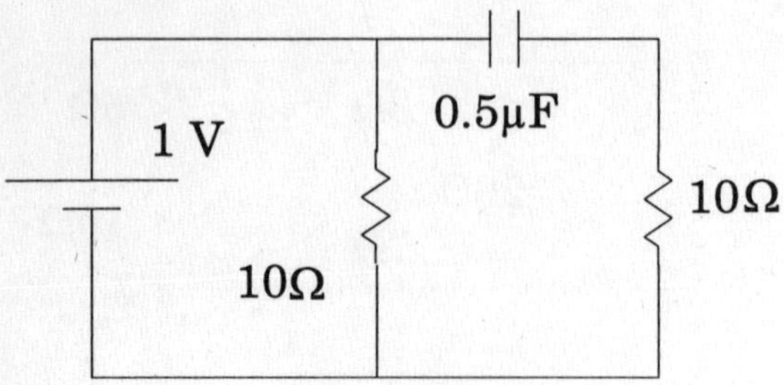

42. What is the steady-state power dissipated by the circuit shown above?

(A) 0.1 W
(B) 0.15 W
(C) 0.2 W
(D) 1 W
(E) 1.5 W

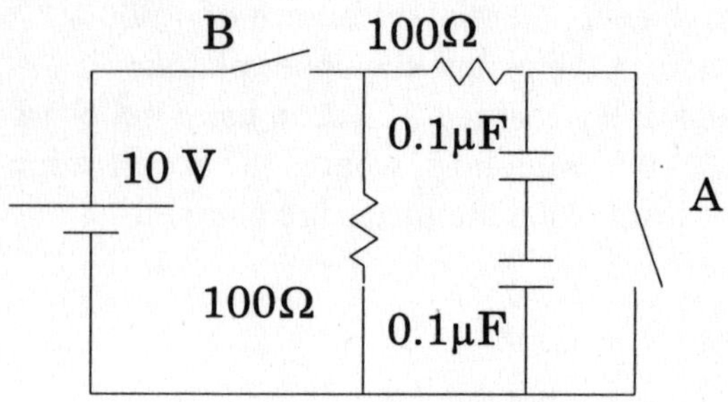

43. Switch *A* is opened and switch *B* is closed in the circuit above. How much charge is stored on each capacitor?

(A) None
(B) 0.1 μC
(C) 0.2 μC
(D) 0.5 μC
(E) 1 μC

44. An electron is moving at constant speed perpendicular to a uniform magnetic field. The trajectory of the electron is a

(A) straight line.
(B) parabola.
(C) circle.
(D) hyperbola.
(E) ellipse.

Questions 45 and 46 refer to the following figure.

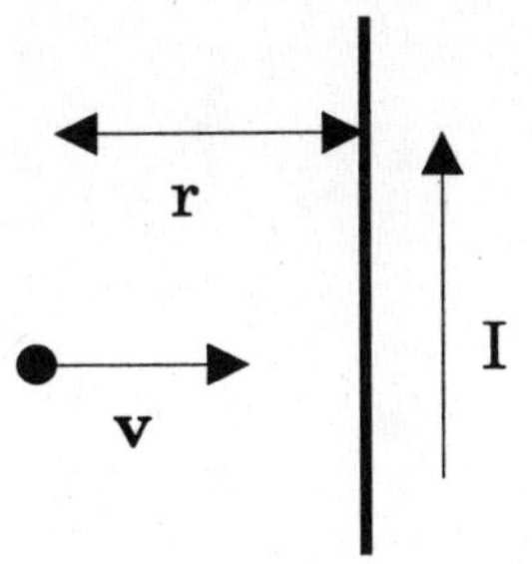

45. A long wire, carrying a current I, produces a magnetic field B at a distance r from a long, straight wire (see figure above). At a distance $2r$, the field is

(A) $4B$.
(B) $2B$.
(C) B.
(D) $\frac{B}{2}$.
(E) $\frac{B}{4}$.

46. A positive charge has a velocity towards the wire, as shown in the figure above. The magnetic force is

(A) zero.
(B) down.
(C) up.
(D) left.
(E) right.

47. A current-carrying wire is in the same plane as a conducting loop. When the current in the wire is increased at a constant rate, the current in the loop

(A) is 0.
(B) increases linearly.
(C) increases quadratically.
(D) is constant.
(E) decreases.

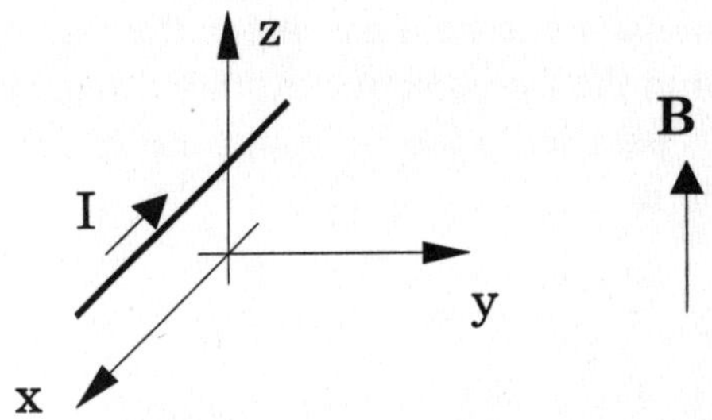

48. A wire is aligned parallel to the x-axis with current flowing as shown in the figure above. A uniform magnetic field points in the $+z$ direction. The force on the wire is

(A) 0.
(B) in the $+y$ direction.
(C) in the $-y$ direction.
(D) in the $+x$ direction.
(E) in the $-x$ direction.

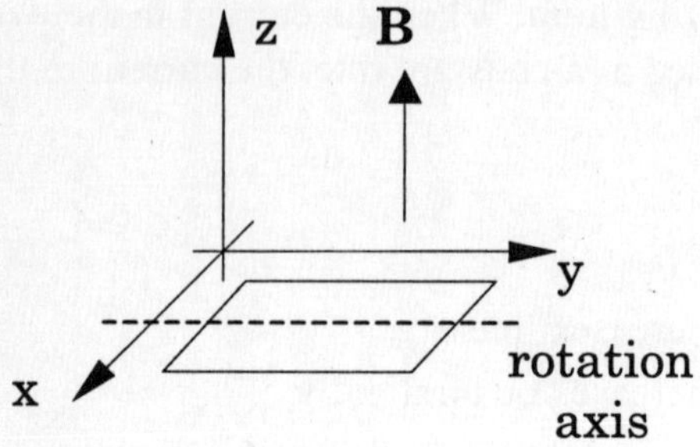

49. A rectangular, conducting coil is rotating about an axis parallel to the y-axis in the presence of a uniform magnetic field parallel to the z-axis, as shown in the figure above. At the instant that the coil is parallel to the x-y plane, the current in the coil is

(A) 0.
(B) clockwise as viewed from above.
(C) counter-clockwise as viewed from above.
(D) clockwise or counter-clockwise depending on the direction of rotation.
(E) not determined by the information given.

50. A wave is traveling across a surface of water at 5 m/s. What is the wavelength if the period is 2 sec?

(A) 2 m
(B) 2.5 m
(C) 5 m
(D) 10 m
(E) 20 m

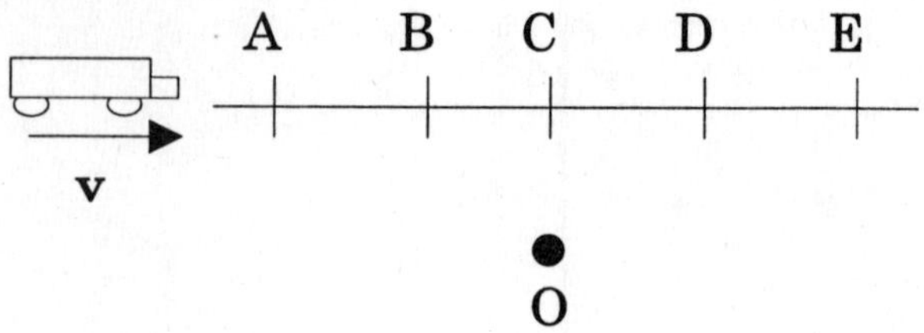

51. A car is traveling past an observer, O, while blowing its horn. At which point is the horn's frequency lowest for the observer?

(A) A
(B) B
(C) C
(D) D
(E) E

52. Two sources, spaced at 2m apart, emit sound of equal amplitude. Their frequencies are 800 Hz and 900 Hz, respectively. Assume the speed of sound is 300 m/s. The nulls (nodes) of the interference pattern

(A) are stationary and spaced at $\frac{1}{3}$ m.
(B) are stationary and spaced at $\frac{3}{8}$ m.
(C) are stationary and spaced at $\frac{1}{2}$ m.
(D) move towards the 800 Hz source.
(E) do not exist.

53. Light of 500 nm wavelength from a distant source impinges on a screen with two narrow slits spaced at 100 μm. How far is the first null in the interference pattern from the central maximum on a surface 1 m from the screen?

(A) 2.5 mm
(B) 5 mm
(C) 1 cm
(D) 2.5 cm
(E) 5 cm

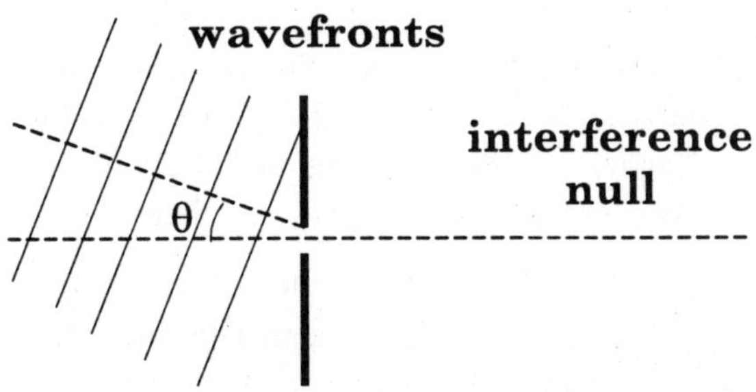

54. A plane wave of 400 nm light is incident on a 25 μm slit in a screen, as shown in the figure above. At what incident angle will the first null of the interference pattern be on a line perpendicular to the screen?

(A) 0.008 rad
(B) 0.016 rad
(C) 0.08 rad
(D) 0.16 rad
(E) No incident angle would create a null on axis.

55. Light from a laser is split into two beams, which are then recombined at a small relative angle. If the laser is tuned to a higher frequency, the resulting interference fringes

(A) remain unchanged.
(B) become less closely spaced.
(C) become more closely spaced.
(D) rotate to a different angle.
(E) disappear.

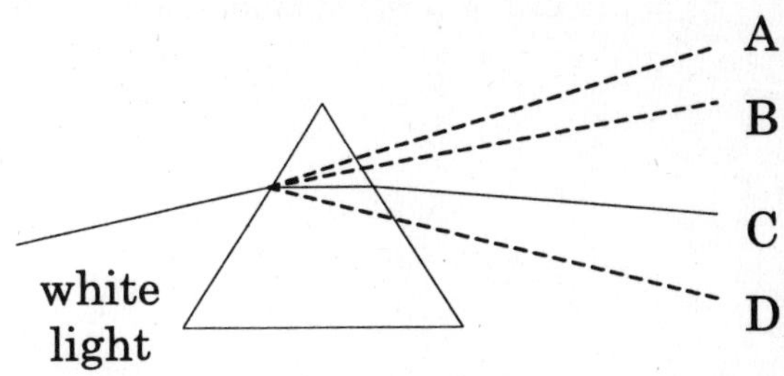

56. White light is incident on a glass prism as shown in the figure above. The refractive index of the glass increases slightly with higher frequency. If red light appears at *C*, blue light appears at

(A) *A*.
(B) *B*.
(C) *C*.
(D) *D*.
(E) cannot be determined from the information given

57. An object is at a distance $\frac{f}{2}$ from a lens of focal length $f > 0$. This lens

(A) forms a real image on the same side of the lens as the object.
(B) forms a virtual image on the same side of the lens as the object.
(C) forms a real image on the opposite side of the lens as the object.
(D) forms a virtual image on the opposite side of the lens as the object.
(E) does not form an image of the object.

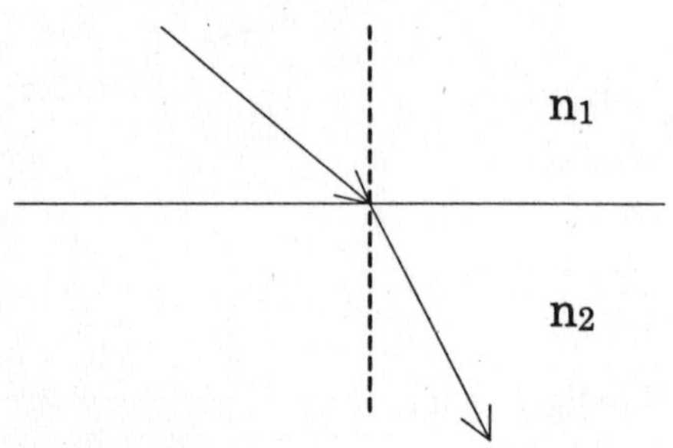

58. The figure above depicts the propagation of light from a medium of refractive index n_1 to one of index n_2. The picture is qualitatively correct

(A) if $n_1 > n_2$.
(B) if $n_1 = n_2$.
(C) if $n_1 < n_2$.
(D) regardless of the relationship between n_1 and n_2.
(E) for no values of n_1 and n_2.

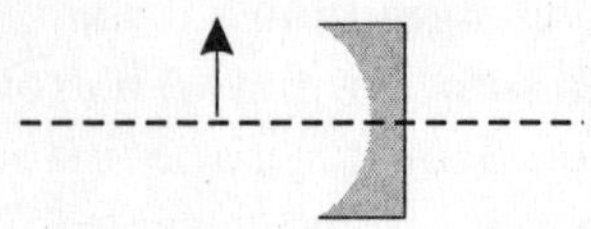

59. An object is placed at a distance equal to the radius of curvature of a concave mirror, as shown above. The image of the object

(A) is real and inverted.
(B) is real and upright.
(C) is virtual and inverted.
(D) is virtual and upright.
(E) does not form.

60. In the Rutherford experiment, alpha particles scatter from a thin gold foil. A few of the particles scatter at large angles away from the incident direction. This is evidence that

(A) gold has electrons.
(B) electrons have a smaller mass than protons.
(C) positive charge is uniformly distributed in gold.
(D) positive charge is concentrated in small, massive particles.
(E) gold is electrically neutral.

61. A red photon is incident on a metal electrode, resulting in the emission of an electron. Which of the following is most likely to occur when a blue photon is incident on the electrode?

(A) emission of two electrons
(B) emission of a proton
(C) emission of a red photon
(D) emission of an electron of lower energy
(E) emission of an electron of higher energy

62. A photon scatters from an atom, resulting in an ejected electron. Compared to the incident photon, the scattered photon has a

(A) higher frequency.
(B) lower frequency.
(C) higher speed.
(D) lower speed.
(E) shorter wavelength.

For questions 63 and 64, the energy levels of a hydrogen atom are represented by the expression

$$E_n = \frac{-E_o}{n^2}.$$

63. If a hydrogen atom is in its lowest energy state, what is the lowest-frequency photon that can be absorbed by the atom while the atom remains in its lowest energy state?

(A) $\frac{E_o}{h}$

(B) $\frac{-E_o}{h}$

(C) $\frac{E_o}{2h}$

(D) $\frac{3E_o}{4h}$

(E) $\frac{3E_o}{8h}$

64. Light of frequency $\frac{E_o}{(2h)}$ is incident on a gas of hydrogen atoms in the ground state. The light

(A) is absorbed, promoting electrons to the $n = 2$ level.
(B) is absorbed, promoting electrons to the $n = 3$ level.
(C) is absorbed, promoting electrons above the $n = 3$ level.
(D) is absorbed, ionizing the hydrogen.
(E) passes through the gas without being absorbed.

Questions 65 and 66 refer to the following description and picture. A beam of electrons is aimed at a pair of slits, producing an interference pattern on a photographic plate, as shown in the figure below.

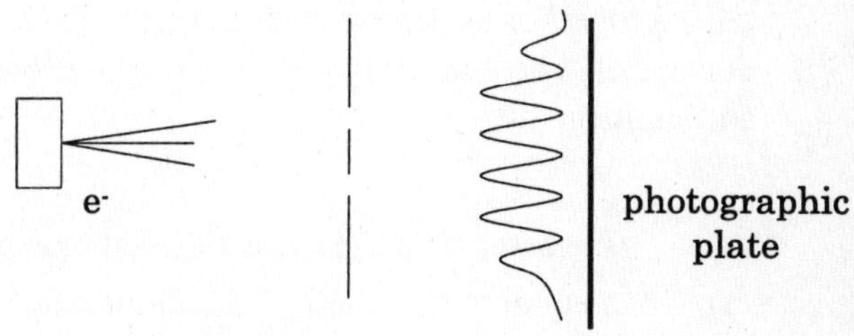

65. When the upper slit is covered, the interference pattern

(A) disappears.
(B) shifts down.
(C) shifts up.
(D) becomes more finely spaced.
(E) remains unchanged.

66. When the electron beam intensity is reduced such that there is at most one electron in the apparatus at once, the interference pattern

(A) disappears.
(B) shifts down.
(C) shifts up.
(D) becomes more finely spaced.
(E) remains unchanged.

67. The half-life of ^{239}Pu is about 25,000 years. After 75,000 years, how much of 1 kg ^{39}Pu remains?

(A) 33 g
(B) 75 g
(C) 125 g
(D) 250 g
(E) 333 g

68. A neutron and a proton collide and remain bound together as a deuterium nucleus ^{2}H (a deuteron). The sum of the masses of the proton and neutron

(A) is greater than the mass of the deuteron.
(B) is equal to the mass of the deuteron.
(C) is less than the mass of the deuteron.
(D) may be greater or less than the mass of the deuteron, but not equal.
(E) cannot be related to the mass of the deuteron.

69. When ^{27}Mg decays by emitting an electron (beta decay), it becomes

(A) ^{26}Mg.
(B) ^{28}Mg.
(C) ^{26}Al.
(D) ^{27}Al.
(E) ^{28}Al.

70. The two isotopes of oxygen ^{16}O and ^{18}O differ by

(A) two protons.
(B) two neutrons.
(C) one proton and one neutron.
(D) two electrons.
(E) one electron and one proton.

SECTION II—FREE RESPONSE

Directions: Answer all six questions. Each question is designed to take approximately 15 minutes to answer. Note that each part within a question may not have equal weight. Show all work to obtain full credit (Assume $g = 10\text{m/s}^2$.)

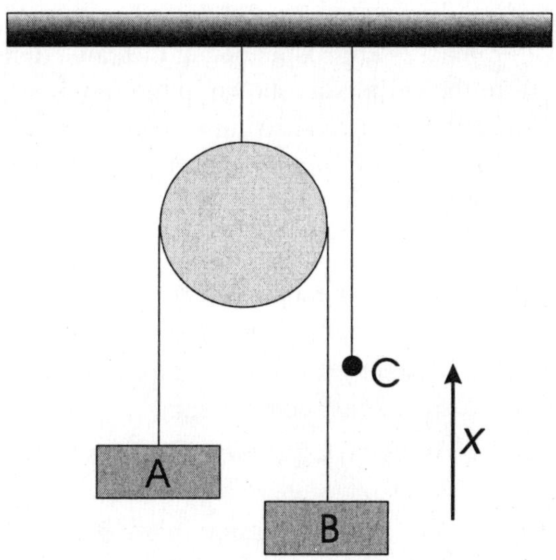

1. The figure above depicts a frictionless Atwood machine with two blocks (A and B) of mass M. A small ball (C) of mass m is suspended above block B by a string. An initial speed V_0 is imparted to the Atwood machine, such that block B rises to collide inelastically with mass C. The positive x-axis is defined as shown. Neglect friction and the masses of the strings connecting the masses.

(a) Find the speed V_1 of the system immediately after the collision of B and C.

(b) Write an equation for the distance, h, above the collision point as a function of time.

(c) How long after the collision does B rise?

(d) Describe the subsequent motion of the Atwood machine, assuming the string remains intact.

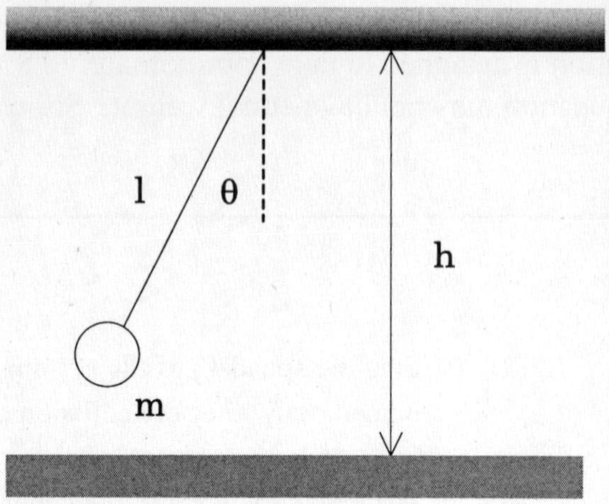

2. A mass m is swinging from a string of length l and is suspended at height h above the floor. The maximum angle of oscillation is θ_0.

(a) Find the speed of the mass as a function of angle, θ.

(b) The string is cut when the mass is at its lowest point, $\theta = 0$. Describe the motion of the mass until it hits the floor, including its position and velocity.

(c) Instead, the string is cut at $\theta = \theta_0$. describe the motion as in part (b) of this question.

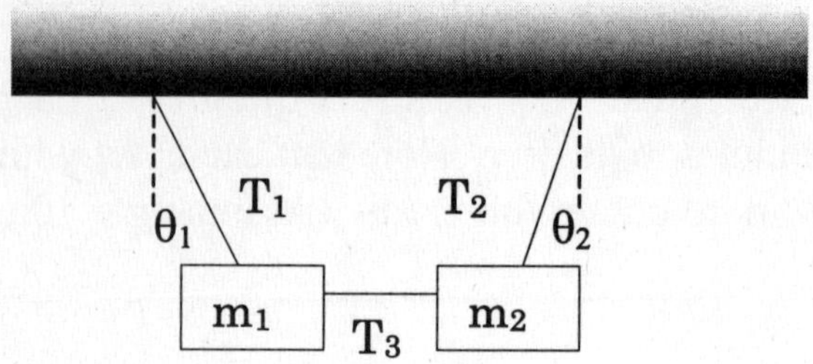

3. Two masses are suspended at the same distance from the ceiling, as shown in the figure above. The tension of each string is denoted by T_1, T_2, and T_3, respectively. The string connecting the two masses is horizontal.

(a) Write an expression for the vertical component of the forces on each mass.

(b) Repeat part (a) of this question for the horizontal components.

(c) Write an expression for T_3 in terms of m_1, θ_1, and g.

(d) Write an expression relating θ_1 to θ_2 in terms of m_1, m_2, and g.

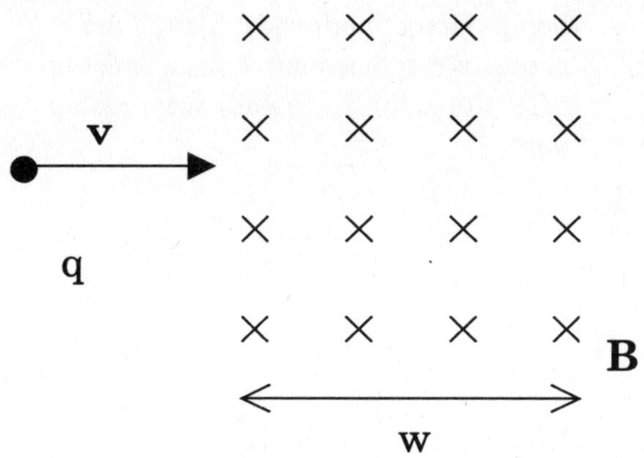

4. A mass m with charge q enters a square region of width w with a uniform magnetic field B pointing into the paper, as shown in the figure above. The point of entry is in the center of the left side.

(a) Describe the trajectory of the mass while it is in the magnetic field region.

(b) Find the speed of the mass when it exits the region.

(c) Find the smallest magnetic field (in terms of m, v, q, and w) that will make the mass exit the square region through the same side that it entered.

(d) Find the magnitude and direction of the electric field required to make the mass's trajectory a straight line.

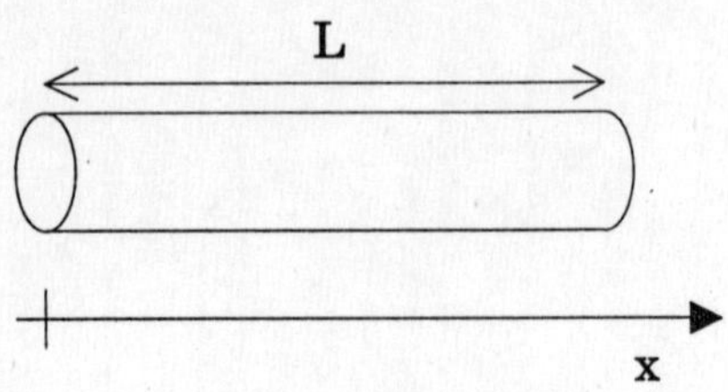

5. Consider a rigid, hollow tube of length L that is open at both ends. Standing waves are created in the tube.

(a) Sketch the acoustic displacement, d, as a function of position for the lowest-order resonance on the graph below.

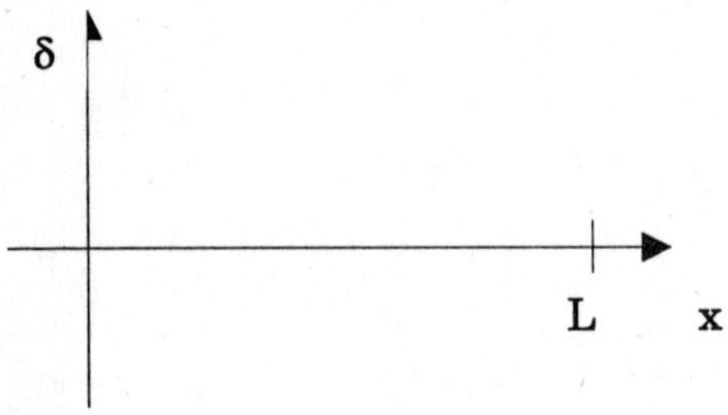

(b) Sketch the acoustic pressure, p, for the lowest-order resonance on the graph below.

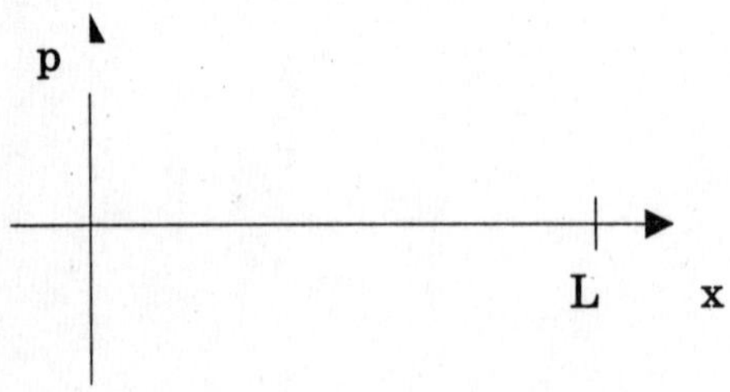

(c) The tube is closed at $x = 0$, instead of being open at both ends. Sketch the acoustic displacement, δ, as a function of position for the lowest-order resonance.

(d) Sketch the acoustic displacement, d, as a function of position for the next lowest resonance.

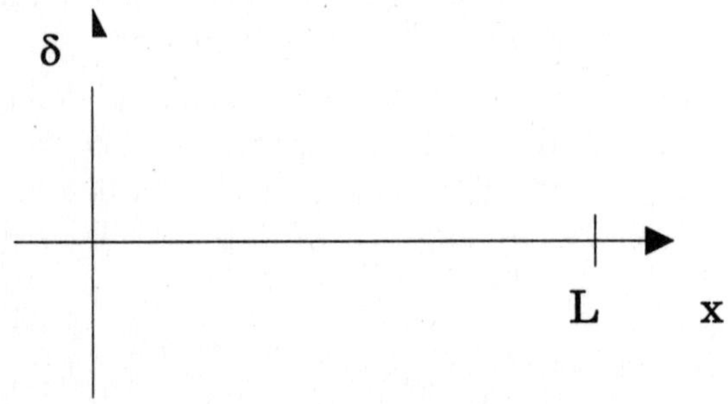

6. An electron and a positron (the electron's antiparticle) are in a circular orbit of radius r about a point halfway between them, as shown in the following figure.

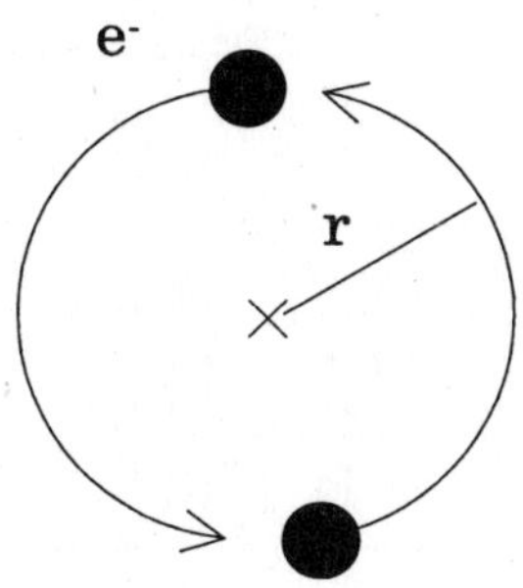

The mass of the positron is equal to that of the electron; the charge has the same magnitude but opposite sign.

(a) If their orbital speed is 10^4 m/s, what is r?

(b) What is the potential energy of the system?

(c) Eventually, the particles annihilate and are replaced by two identical photons. What is their wavelength? Neglect the mechanical energy of the electron/positron pair.

(d) How does the energy of the photons compare to the potential energy computed in part (b) of this question?

Physics B Practice Test 2

ANSWERS AND EXPLANATIONS

MULTIPLE CHOICE SECTION

QUICK-SCORE ANSWERS									
1.E	8.B	15.C	22.E	29.A	36.B	43.D	50.D	57.B	64.E
2.B	9.D	16.A	23.C	30.B	37.C	44.C	51.E	58.C	65.A
3.D	10.B	17.B	24.D	31.D	38.C	45.D	52.D	59.A	66.E
4.Z	11.C	18.D	25.B	32.A	39.D	46.B	53.A	60.D	67.C
5.B	12.C	19.B	26.D	33.D	40.C	47.D	54.B	61.E	68.A
6.D	13.A	20.C	27.C	34.E	41.D	48.B	55.C	62.B	69.D
7.C	14.D	21.D	28.B	35.B	42.A	49.A	56.D	63.D	70.B

1. **The correct answer is (E).** Momentum and total energy are always conserved. Kinetic energy is only conserved in elastic collisions.

2. **The correct answer is (B).** The work done by friction is equal to the initial kinetic energy of the object: $\mu\ mgx = \frac{1}{2}mv^2$, so

$$\mu = \frac{v^2}{(2gx)} = \frac{5^2}{(2 \cdot 10 \cdot 5)} = 0.25 \cdot$$

3. **The correct answer is (D).** The acceleration vector points to the left, so the velocity vector will decrease in length until it points to the left.

4. **The correct answer is (A).** The speed is proportional to *t*, while the displacement is proportional to t^2. So, *v* is proportional to $\sqrt{s}$.

5. **The correct answer is (B).** Momentum is conserved, so $m\Delta v = (M + 2m)V \cdot$
   Set the kinetic of the pendulum after collision equal to the potential energy at the highest point: $\frac{1}{2}(M + 2m)V^2 = (M + 2m)gh$. Solve for the difference in speed: $\Delta v = \frac{2m + M}{m}\sqrt{2gh}$.

6. **The correct answer is (D).** The time required to fall to the ground is $\sqrt{\left(125 * \frac{2}{10}\right)} = 5\,\text{sec}$. When this is substituted in the equation for the horizontal motion, the distance, $d = vt = 200*5 = 1000$ m, is obtained.

7. **The correct answer is (C).** When the center of mass is immediately above the support point, the box is at the edge of stability. Simple trigonometry relates the sides of the box to the tangent of the tipping angle.

8. **The correct answer is (B).** The resultant force is the vector sum of the forces given. Its magnitude may be found to be 5 N by using the Pythagorean theorem. To find the acceleration, divide the force by the mass: $\frac{5}{10} = 0.5\text{m/s}^2$.

9. **The correct answer is (D).** Momentum is conserved, so the momentum of the boat is equal and opposite to Jack's: (80kg)(3m/s)=240kg-m/s. Divide by the mass of the boat to find its speed: $\frac{240}{20} = 12\text{m/s}$.

10. **The correct answer is (B).** The acceleration is always a constant, g, the acceleration of gravity. The velocity is zero as the ball changes direction.

11. **The correct answer is (C).** The change in momentum is equal to the product of the force and the time over which it acts. The speed of the ball is equal after bouncing from the wall, but its direction is opposite, so the change in velocity is 20m/s. The change in momentum is (0.04kg)(20m/s) = 0.8kg–m/s. Divide the momentum change by the force to find the time: (0.8kg–m/s)/(16N) = 0.05s = 50ms.

12. **The correct answer is (C).** Work is force times net displacement. The force required is the weight of the box, 200N. The work is (200N)(15m) = 3000J.

13. **The correct answer is (A).** Since the displacement is zero, so is the work.

14. **The correct answer is (D).** The velocity is not constant because its direction is changing, but its magnitude (speed) is constant. Likewise, the direction of the linear acceleration and force are changing.

15. **The correct answer is (C).** The net force is the mass times the acceleration: $(15\text{kg})(0.3\text{m/s}^2) = 4.5\text{N}$. The force of friction is $(0.2)(15\text{kg})(10\text{m/s}^2) = 30\text{N}$. Barbara must supply the sum of these two: $4.5 + 30 = 34.5\text{N}$.

16. **The correct answer is (A).** The weight is $(20\text{kg})(10\text{m/s}^2) = 200\text{N}$. Subtract this from 250N to find the net force of 50N. Divide by the mass to find the acceleration: $\dfrac{50\text{N}}{20\text{kg}} = 2.5\text{m/s}^2$.

17. **The correct answer is (B).** The upward force the scale reads 60kg and exerts when the elevator is at rest is $(60\text{kg})(10\text{m/s}^2) = 600\text{N}$. When the acceleration of the elevator is 1 m/s^2, there is a net downward force of $(60\text{kg})(1\text{m/s}^2) = 60\text{N}$, which means that the scale only exerts $600\text{N}-60\text{N} = 540\text{N}$. Thus, the scale reads $\dfrac{540\text{N}}{10\text{m/s}^2} = 54\text{kg}$.

18. **The correct answer is (D).** The period of a pendulum is inversely proportional to the square root of the gravitational acceleration. The period is greater on Mars by a factor of $\sqrt{\dfrac{1}{0.4}} = 1.6$.

19. **The correct answer is (B).** The force required to keep the car in circular motion is $\dfrac{mv^2}{R} = \dfrac{(200\text{kg})(50\text{m/s}^2)^2}{100m} = 5000\text{N}$.

20. **The correct answer is (C).** The potential energy of the ball is proportional to its height. The total mechanical energy of the ball is equal to the potential energy at the top of the trajectory. The ball loses 20% of its energy in the first bounce, so when it loses another 20% on the next bounce, the height is $(80\text{cm})(0.8) = 64\text{cm}$.

21. **The correct answer is (D).** The force required is $F = \mu N = \mu mg = (0.2)(300\text{kg})(10\text{m/s}^2) = 600\text{N}$. $P = Fv = (600\text{N})(2\text{m/s}) = 1200\text{W}$.

22. **The correct answer is (E).** The momenta are equal and opposite, so the total momentum is zero.

23. **The correct answer is (C).** The mass will bounce up when the potential energy lost by falling is entirely stored in the potential energy of the spring: $mgx = \left(\frac{1}{2}\right)kx^2$. Solve for x to find the distance:

$$x = \frac{2mg}{k} = \frac{2(2\text{kg})(10\text{m/s}^2)}{50N/\text{m}} = 0.8\text{m}.$$

24. **The correct answer is (D).** The period is proportional to the square root of the mass, so the period increases by a factor of $\sqrt{2}$.

25. **The correct answer is (B).** Equate the force of friction to the centripetal force required: $\mu mg = \frac{mv^2}{R} = \frac{m(2\pi Rf)^2}{R} = 4\pi^2 mRf^2$, where $f = 2$ is the rotation frequency. Solve for $R = \frac{\mu g}{(4\pi^2 f^2)}$.

26. **The correct answer is (D).** When the kinetic energy equals the potential energy, the potential energy is half the total energy: $\left(\frac{1}{2}\right)kx^2 = \left(\frac{1}{2}\right)\left(\frac{1}{2}\right)kA^2$, where A is the amplitude of the oscillation. Solve for x in terms of A: $x = \frac{A}{\sqrt{2}} = \frac{14\text{cm}}{\sqrt{2}} \approx 10\text{cm}$.

27. **The correct answer is (C).** While both ice and water are present, the temperature remains at the freezing point. Added heat goes in to melting ice, not to changing the temperature.

28. **The correct answer is (B).** The potential energy lost by the water is converted into heat. The potential energy loss per kilogram is $gh = (10\text{m/s}^2)(60\text{m}) = 600\text{m}^2/\text{s}^2$. Divide by the specific heat of water to find the temperature change: $\frac{600m^2/s^2}{4{,}000\ J/(kg\text{-}°C)} = 0.15°C$.

29. **The correct answer is (A).** The work is proportional to the area under the path. Path A includes the most area.

30. **The correct answer is (B).** A refrigerator does some work, so it must add some heat to that is removed from the interior. Thus, the heat added to the kitchen is greater, not equal to, the heat removed from the interior.

31. **The correct answer is (D).** According to the ideal gas law, the pressure is proportional to the temperature.

32. **The correct answer is (A).** The maximum efficiency is greater when the temperature difference is greater.

33. **The correct answer is (D).** When the gas is compressed, the pressure increases, according to the ideal gas law. In addition, since work is done on the gas, the internal energy increases, thereby increasing the temperature.

34. **The correct answer is (E).** The work is independent of the path since the electric force is conservative.

35. **The correct answer is (B).** The electric field of the plate is constant and directed away from the plate. The field is in opposite directions on opposite sides of the plate, so the force is equal and oppositely directed.

36. **The correct answer is (B).** The vertical components of the electric field cancel at the location of the upper charge.

37. **The correct answer is (C).** The charge induces an opposite charge on the sphere resulting in an attractive force, regardless of the sign of the charge.

38. **The correct answer is (C).** The electric field is $\frac{(1{,}000V)}{(0.1\text{m})} = 10{,}000\frac{V}{\text{m}}$. The upward electrical force is $(10^{-18}\text{C})(10{,}000\text{V/m}) = 10^{-14}\text{N}$. The gravitational force is $(10^{-15}\text{kg})(10\text{m/s}^2) = 10^{-14}\text{N}$. Since the net force is zero, so is the acceleration.

39. **The correct answer is (D).** The energy required to move the charges is equal to the change in potential energy of the charges.

40. **The correct answer is (C).** The current, I, in each parallel branch of the circuit is inversely proportional to the resistance of that branch. Hence, the central branch has twice the current of the others. The resistances from left to right are 20Ω, 10Ω, and 20Ω, respectively. The power dissipation in a resistor R is I^2R. The 7Ω resistor has the highest value of I^2R.

41. **The correct answer is (D).** The effective resistance is found by combining the parallel resistances of 20Ω, 10Ω, and 20Ω:

$$\frac{1}{R} = \frac{1}{20} + \frac{1}{10} + \frac{1}{20}$$

$$R = 5\Omega$$

$$I = \frac{V}{R} = \frac{10volts}{5\Omega} = 2A$$

42. **The correct answer is (A).** No DC current flows through the capacitor, so current only flows through the 10Ω resistor. The power dissipated in the resistor is

$$P = IV = \frac{V^2}{R} = \frac{1}{10} = 0.1W.$$

43. **The correct answer is (D).** Since the capacitors are in series, the 10V is equally divided between them. So, the charge on each one is $Q = CV = (0.1\mu\text{F})(5\text{V}) = 0.5\mu\text{C}$.

44. **The correct answer is (C).** The magnetic force is always perpendicular to the velocity, so the electron is in uniform circular motion.

45. **The correct answer is (D).** The magnetic field near a long wire is inversely proportional to the distance from the wire.

46. **The correct answer is (B).** The magnetic field is out of the paper at the location of the charge, so the force on the charge is down, according to the right-hand rule.

47. **The correct answer is (D).** According to Faraday's Law, the current in the loop is proportional to the rate of change of the magnetic flux. Since the current in the wire increases at a constant rate, the flux does also. Thus, the current is constant.

48. **The correct answer is (B).** According to the right-hand rule, the force is to the right ($+y$).

49. **The correct answer is (A).** The rate of change of the flux is zero when the coil is in the x-y plane. According to Faraday's Law, the emf is zero.

50. **The correct answer is (D).** The frequency is the reciprocal of the period: $\frac{1}{(2s)} = 0.5\text{Hz}$. The wavelength is equal to the speed divided by the frequency: $\frac{(5\text{m/s})}{0.5\text{hz}} = 10\text{m}$.

51. **The correct answer is (E).** The frequency is lowest when the component of velocity away from the observer is the larger. Only choices (D) and (E) have components away from the observer. Choice (E) has a larger component.

52. **The correct answer is (D).** Since the frequencies of the sources are different, there are no stationary nulls in the pattern. There are nulls, however, where the waves cancel at any moment.

53. **The correct answer is (A).** The diffraction angle is given by

$$\sin\theta = \frac{\lambda}{2d}$$

$$= \frac{500nm}{2 \cdot 100\mu m} = \frac{5\times10^{-7}m}{2\times10^{-4}m} = 2.5\times10^{-3}$$

$$\theta \approx 2.5mrad$$

At 1 m, the distance is $(2.5mrad)(1\text{m}) = 2.5\text{mm}$.

54. **The correct answer is (B).** The first null in single-slit interference pattern is given by

$$\sin\theta = \frac{\lambda}{d}$$

$$= \frac{400nm}{25\mu m} = \frac{4\times10^{-7}m}{2.5\times10^{-5}m} = 0.016$$

$$\theta \approx 0.016rad$$

If the angle of incidence is θ, the null will be on the axis.

55. **The correct answer is (C).** When the frequency is higher, the wavelength is shorter, so the interference pattern is more closely spaced.

56. **The correct answer is (D).** The higher the frequency, the greater the angular deflection. Blue light has a higher frequency than red light.

57. **The correct answer is (B).** Since the object is closer than one focal length from the lens, the image is virtual and on the same side of the lens.

58. **The correct answer is (C).** According to Snell's Law, the angle of refraction is smaller if the index is higher.

59. **The correct answer is (A).** The radius of curvature is twice the focal length of the mirror. So, the mirror forms a real, inverted image.

60. **The correct answer is (D).** The large scattering angles mean that electrons must be much lighter than the positive particles in gold. The fact that only a few of them scatter through large angles means that the positive particles are small.

61. **The correct answer is (E).** Blue light has a higher frequency (hence, higher energy photons) than red light. In the photoelectric effect, higher frequency light causes the emission of more energetic electrons.

62. **The correct answer is (B).** Some of the incident photon's energy is given up to release the electron, so the scatter photon has a lower energy. The energy of a photon is proportional to its frequency.

63. **The correct answer is (D).** The smallest increase in energy allowed for the atom in its lowest state is $E_2 - E_1 = E_o\left(-\frac{1}{4}+1\right)=\left(\frac{3}{4}\right)E_o$. The frequency is given by the energy divided by Planck's constant, *h*.

64. **The correct answer is (E).** The photons do not have enough energy to excite the atom, so they are not absorbed.

65. **The correct answer is (A).** If the electrons cannot pass through both slits, there is no interference pattern.

66. **The correct answer is (E).** The intensity of the electron beam does not change the character of the interference pattern.

67. **The correct answer is (C).** After *n* half-lives, the fraction of the original sample that remains is $\left(\frac{1}{2}\right)^n$. Since 75,000 years corresponds to three half-lives, $\frac{1}{8}$ of 1kg remains.

68. **The correct answer is (A).** Since the mass of the deuterium includes (negative) binding energy, the mass is less than the sum of the constituents.

69. **The correct answer is (D).** The atomic mass does not change in beta decay.

70. **The correct answer is (B).** The number of protons is the same for all isotopes of oxygen. The difference in mass is caused by two neutrons.

SECTION II—FREE RESPONSE

1. **(a)** Linear momentum is conserved:

$$2MV_0 = (2M + m)V_1$$

$$V_1 = \frac{2MV_0}{(2M + m)}$$

(b) Immediately after collision, the net force on the system is mg.

Hence, the system experiences an acceleration of: $a = \frac{mg}{(2M + m)}$.

$$h = V_1 t - \tfrac{1}{2}at^2$$

The height above the collision point is $h = V_1 t - \frac{mg}{2(2M + m)}t^2$.

(c) Mass B continues to rise until the velocity is zero:

$V = V_1 - at = 0$

$V_1 = at$

$$t = \frac{V_1}{a}$$

$$= \frac{V_1}{\left[\frac{mg}{(2m + m)}\right]} = \frac{(2M + m)}{mg}V_1 = \frac{(2M + m)}{mg} \cdot \frac{2MV_0}{(2M + m)} = \frac{2MV_0}{mg}.$$

(d) After reversing direction at the time found in part (c) of this question, mass B continues to move down, accelerating until the string lifts mass C off of B. At this point, the velocity remains constant at V_0.

2. **(a)** Use conservation of energy to find the relationship between θ and speed. Write the expression for energy as a function of θ:

$$\tfrac{1}{2}mv^2 + mgl(1 - \cos\theta) = mgl(1 - \cos\theta_0)$$

Solve for v:

$$\tfrac{1}{2}v^2 = gl(\cos\theta - \cos\theta_0)$$

$$v(\theta) = \sqrt{2gl(\cos\theta - \cos\theta_0)}$$

(b) When θ = 0, the speed is $v_x = v(0) = \sqrt{2gl(1 - \cos\theta_0)}$.

This horizontal component remains constant after the string is cut.

The vertical motion is uniformly accelerated: $h - l = \frac{1}{2}gt^2$.

The mass follows a parabolic trajectory to the floor. It reaches the floor at

$t = \sqrt{\frac{2(h-l)}{g}}$. The horizontal displacement is

$$x = v_x t$$
$$= 2\sqrt{l(h-l)(1-\cos\theta_0)}$$

(c) The velocity is zero when $q = q_0$. The mass falls straight down. Its initial height above the floor is $h - l\cos\theta_0$. The time required is

$$h - l\cos\theta_0 = \frac{1}{2}gt^2$$

given by $t = \sqrt{\frac{2(h - l\cos\theta_0)}{g}}$

3. **(a)** The vertical component of the tension on each mass must balance the gravitational force:

$$T_1\cos\theta_1 = m_1 g$$
$$T_2\cos\theta_2 = m_2 g$$

(b) The horizontal component of the tension in the strings must add to zero for each mass: $T_1\sin\theta_1 = T_2\sin\theta_2 = T_3$.

(c) The results of parts (a) and (b) of this question can be combined to eliminate T_1: $T_3 = m_1 g\tan\theta_1$.

(d) A similar result as found in part (c) of this question also holds true for mass 2, thus $T_3 = m_1 g\tan\theta_1 = m_2 g\tan\theta_2$. Therefore, $\frac{\tan\theta_1}{\tan\theta_2} = \frac{m_2}{m_1}$

4. **(a)** The trajectory is a part of a circle because the force is constant and perpendicular to the velocity.

(b) The magnetic field does no work on the mass for the reason cited in part (a) of this question. The speed remains constant, v.

(c) The radius of curvature must be less than or equal to $\frac{w}{4}$ for the mass to exit through the same side. The radius of the trajectory is related to the velocity by applying the force law:

$$qvB = \frac{mv^2}{r}$$

$$r = \frac{mv}{qB} \leq \frac{w}{4}$$

Solving for B:

$$B \geq \frac{4mv}{qw}$$

(d) The electric force cancels magnetic force if $qE = qvB$ or $E = vB$

5. **(a)** There are displacement antinodes at the open ends. Either the dashed or solid line is correct:

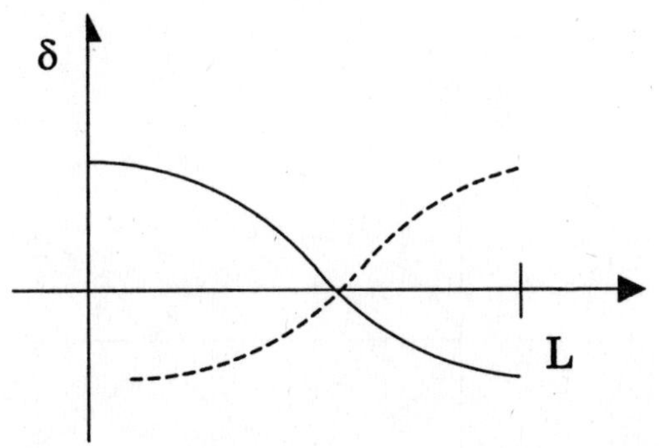

(b) The pressure has nodes at the open ends:

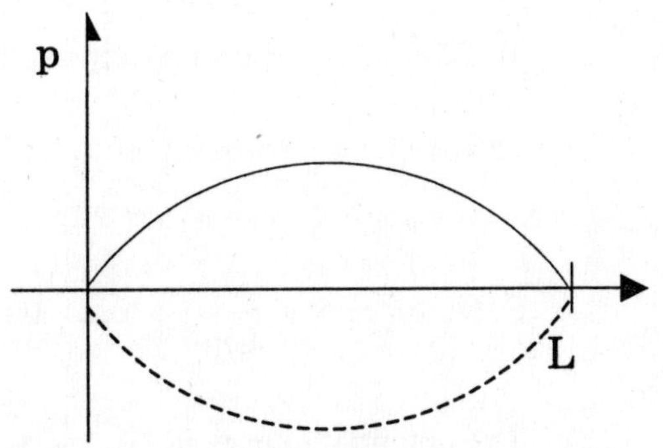

(c) There is a displacement node at the closed end and an antinode at the open end:

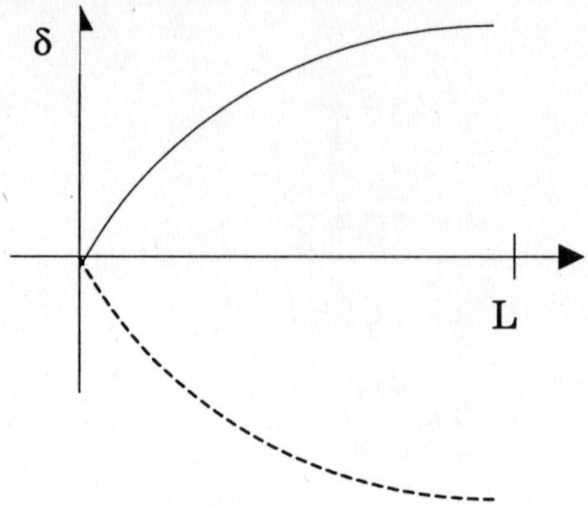

(d) The next resonance also has a displacement node at the closed end and an antinode at the open end:

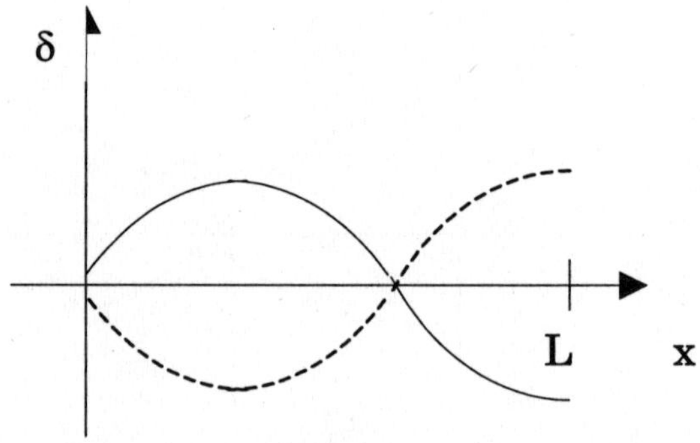

6. **(a)** To find the relationship between the radius of the orbit and the speed, set the electric force equal to the centripetal force: $\frac{mv^2}{r} = k\frac{e^2}{(2r)^2}$.

Solve for r: $r = \frac{ke^2}{4mv^2} = 6.3\times10^{-7}\,\text{m} = 630\text{nm}$.

(b) The potential energy is $k\frac{e^2}{2r} = 1.8\times10^{-22}\,J$.

(c) The total energy released is $2mc^2$; each photon has energy mc^2. The frequency is

$\nu = \frac{mc^2}{h}$. Therefore the wavelength

is $\lambda = \frac{c}{\nu} = \frac{h}{mc} = 2.4\times10^{-12}\,\text{m} = 2.4\,pm$.

(d) The energy of each photon is $mc^2 = 8.2\times10^{-14}\,J$.
This is much greater than the potential energy found in part (b) of this question.

ANSWER SHEET FOR PHYSICS C PRACTICE TEST I

Section I: Mechanics

1 ⊂A⊃ ⊂B⊃ ⊂C⊃ ⊂D⊃ ⊂E⊃
2 ⊂A⊃ ⊂B⊃ ⊂C⊃ ⊂D⊃ ⊂E⊃
3 ⊂A⊃ ⊂B⊃ ⊂C⊃ ⊂D⊃ ⊂E⊃
4 ⊂A⊃ ⊂B⊃ ⊂C⊃ ⊂D⊃ ⊂E⊃
5 ⊂A⊃ ⊂B⊃ ⊂C⊃ ⊂D⊃ ⊂E⊃
6 ⊂A⊃ ⊂B⊃ ⊂C⊃ ⊂D⊃ ⊂E⊃
7 ⊂A⊃ ⊂B⊃ ⊂C⊃ ⊂D⊃ ⊂E⊃
8 ⊂A⊃ ⊂B⊃ ⊂C⊃ ⊂D⊃ ⊂E⊃
9 ⊂A⊃ ⊂B⊃ ⊂C⊃ ⊂D⊃ ⊂E⊃
10 ⊂A⊃ ⊂B⊃ ⊂C⊃ ⊂D⊃ ⊂E⊃
11 ⊂A⊃ ⊂B⊃ ⊂C⊃ ⊂D⊃ ⊂E⊃
12 ⊂A⊃ ⊂B⊃ ⊂C⊃ ⊂D⊃ ⊂E⊃
13 ⊂A⊃ ⊂B⊃ ⊂C⊃ ⊂D⊃ ⊂E⊃
14 ⊂A⊃ ⊂B⊃ ⊂C⊃ ⊂D⊃ ⊂E⊃
15 ⊂A⊃ ⊂B⊃ ⊂C⊃ ⊂D⊃ ⊂E⊃
16 ⊂A⊃ ⊂B⊃ ⊂C⊃ ⊂D⊃ ⊂E⊃
17 ⊂A⊃ ⊂B⊃ ⊂C⊃ ⊂D⊃ ⊂E⊃
18 ⊂A⊃ ⊂B⊃ ⊂C⊃ ⊂D⊃ ⊂E⊃
19 ⊂A⊃ ⊂B⊃ ⊂C⊃ ⊂D⊃ ⊂E⊃
20 ⊂A⊃ ⊂B⊃ ⊂C⊃ ⊂D⊃ ⊂E⊃
21 ⊂A⊃ ⊂B⊃ ⊂C⊃ ⊂D⊃ ⊂E⊃
22 ⊂A⊃ ⊂B⊃ ⊂C⊃ ⊂D⊃ ⊂E⊃
23 ⊂A⊃ ⊂B⊃ ⊂C⊃ ⊂D⊃ ⊂E⊃
24 ⊂A⊃ ⊂B⊃ ⊂C⊃ ⊂D⊃ ⊂E⊃
25 ⊂A⊃ ⊂B⊃ ⊂C⊃ ⊂D⊃ ⊂E⊃
26 ⊂A⊃ ⊂B⊃ ⊂C⊃ ⊂D⊃ ⊂E⊃
27 ⊂A⊃ ⊂B⊃ ⊂C⊃ ⊂D⊃ ⊂E⊃
28 ⊂A⊃ ⊂B⊃ ⊂C⊃ ⊂D⊃ ⊂E⊃
29 ⊂A⊃ ⊂B⊃ ⊂C⊃ ⊂D⊃ ⊂E⊃
30 ⊂A⊃ ⊂B⊃ ⊂C⊃ ⊂D⊃ ⊂E⊃
31 ⊂A⊃ ⊂B⊃ ⊂C⊃ ⊂D⊃ ⊂E⊃
32 ⊂A⊃ ⊂B⊃ ⊂C⊃ ⊂D⊃ ⊂E⊃
33 ⊂A⊃ ⊂B⊃ ⊂C⊃ ⊂D⊃ ⊂E⊃
34 ⊂A⊃ ⊂B⊃ ⊂C⊃ ⊂D⊃ ⊂E⊃
35 ⊂A⊃ ⊂B⊃ ⊂C⊃ ⊂D⊃ ⊂E⊃

Section I: Electricity and Magnetism

36 ⊂A⊃ ⊂B⊃ ⊂C⊃ ⊂D⊃ ⊂E⊃
37 ⊂A⊃ ⊂B⊃ ⊂C⊃ ⊂D⊃ ⊂E⊃
38 ⊂A⊃ ⊂B⊃ ⊂C⊃ ⊂D⊃ ⊂E⊃
39 ⊂A⊃ ⊂B⊃ ⊂C⊃ ⊂D⊃ ⊂E⊃
40 ⊂A⊃ ⊂B⊃ ⊂C⊃ ⊂D⊃ ⊂E⊃
41 ⊂A⊃ ⊂B⊃ ⊂C⊃ ⊂D⊃ ⊂E⊃
42 ⊂A⊃ ⊂B⊃ ⊂C⊃ ⊂D⊃ ⊂E⊃
43 ⊂A⊃ ⊂B⊃ ⊂C⊃ ⊂D⊃ ⊂E⊃
44 ⊂A⊃ ⊂B⊃ ⊂C⊃ ⊂D⊃ ⊂E⊃
45 ⊂A⊃ ⊂B⊃ ⊂C⊃ ⊂D⊃ ⊂E⊃
46 ⊂A⊃ ⊂B⊃ ⊂C⊃ ⊂D⊃ ⊂E⊃
47 ⊂A⊃ ⊂B⊃ ⊂C⊃ ⊂D⊃ ⊂E⊃
48 ⊂A⊃ ⊂B⊃ ⊂C⊃ ⊂D⊃ ⊂E⊃
49 ⊂A⊃ ⊂B⊃ ⊂C⊃ ⊂D⊃ ⊂E⊃
50 ⊂A⊃ ⊂B⊃ ⊂C⊃ ⊂D⊃ ⊂E⊃
51 ⊂A⊃ ⊂B⊃ ⊂C⊃ ⊂D⊃ ⊂E⊃
52 ⊂A⊃ ⊂B⊃ ⊂C⊃ ⊂D⊃ ⊂E⊃
53 ⊂A⊃ ⊂B⊃ ⊂C⊃ ⊂D⊃ ⊂E⊃
54 ⊂A⊃ ⊂B⊃ ⊂C⊃ ⊂D⊃ ⊂E⊃
55 ⊂A⊃ ⊂B⊃ ⊂C⊃ ⊂D⊃ ⊂E⊃
56 ⊂A⊃ ⊂B⊃ ⊂C⊃ ⊂D⊃ ⊂E⊃
57 ⊂A⊃ ⊂B⊃ ⊂C⊃ ⊂D⊃ ⊂E⊃
58 ⊂A⊃ ⊂B⊃ ⊂C⊃ ⊂D⊃ ⊂E⊃
59 ⊂A⊃ ⊂B⊃ ⊂C⊃ ⊂D⊃ ⊂E⊃
60 ⊂A⊃ ⊂B⊃ ⊂C⊃ ⊂D⊃ ⊂E⊃
61 ⊂A⊃ ⊂B⊃ ⊂C⊃ ⊂D⊃ ⊂E⊃
62 ⊂A⊃ ⊂B⊃ ⊂C⊃ ⊂D⊃ ⊂E⊃
63 ⊂A⊃ ⊂B⊃ ⊂C⊃ ⊂D⊃ ⊂E⊃
64 ⊂A⊃ ⊂B⊃ ⊂C⊃ ⊂D⊃ ⊂E⊃
65 ⊂A⊃ ⊂B⊃ ⊂C⊃ ⊂D⊃ ⊂E⊃
66 ⊂A⊃ ⊂B⊃ ⊂C⊃ ⊂D⊃ ⊂E⊃
67 ⊂A⊃ ⊂B⊃ ⊂C⊃ ⊂D⊃ ⊂E⊃
68 ⊂A⊃ ⊂B⊃ ⊂C⊃ ⊂D⊃ ⊂E⊃
69 ⊂A⊃ ⊂B⊃ ⊂C⊃ ⊂D⊃ ⊂E⊃
70 ⊂A⊃ ⊂B⊃ ⊂C⊃ ⊂D⊃ ⊂E⊃

Physics Formulas
TABLE OF INFORMATION

Constants and Conversion Factors

1 unified atomic mass unit.	$1\ u = 1.66 \times 10^{-27}$ kg = 931 MeV / c^2
Proton mass	$m_p = 1.67 \times 10^{-27}$ kg
Neutron mass	$m_n = 1.67 \times 10^{-27}$ kg
Electron mass	$m_e = 9.11 \times 10^{-31}$ kg
Magnitude of electron charge	$e = 1.60 \times 10^{-19}$
Avogadro's number	$N_0 = 6.02 \times 10^{23}$ mol^{-1}
Universal gas constant	R = 8.31 J / (mol × K)
Boltzmann's constant	$k_B = 1.38 \times 10^{-23}$ J/K
Speed of light	$c = 3.00 \times 10^8$ m/s
Planck's constant	$h = 6.63 \times 10^{-34}$J × s = 4.14×10^{-15} eV × s
	$Hc = 1.99 \times 10^{-25}$J × m = 1.24×10^3 eV × nm
Vacuum permittivity	$e_0 = 8.85 \times 10^{-12}$C2/N × m^2
Coulomb's law constant	$k = 1/4pe_0 = 9.0 \times 10^9$N × m^2/C^2
Vacuum permeability	$m_0 = 4p \times 10^{-7}$Wb/(A × m)
Magnetic constant	k¢ = $m_0/4p \times 10^{-7}$Wb/(A × m)
Universal gravitational constant	$G = 6.67 \times 10^{-11}$ m^3/kg × s^2
Acceleration due to gravity at the Earth's surface	g = 9.8 m/s^2
1 atmosphere pressure	1 atm = 1.0×10^5 N/m^2 = 1.0×10^5 Pa
1 electron volt	1 eV = 1.60×10^{-19}J
1 angstrom	1 Å = 1×10^{-10} m

Units

Name	Symbol
meter	m
kilogram	kg
second	s
ampere	A
kelvin	K
mole	mol
hertz	Hz
newton	N
pascal	Pa
joule	J
watt	W
coulomb	C
volt	V
ohm	W
henry	H
farad	F
weber	Wb
tesla	T
degree Celsius	°C
electron-volt	eV

Prefixes

Factor	Prefix	Symbol
10^9	giga	G
10^6	mega	M
10^3	kilo	k
10^{-2}	centi	c
10^{-3}	milli	m
10^{-6}	micro	m
10^{-9}	nano	n
10^{-12}	pico	p

Values of Trigonometric Functions For Common Angles

Angle	Sin	Cos	Tan
0°	0	1	0
30°	1 / 2	$\sqrt{3}/2$	$\sqrt{3}/3$
37°	3 / 5	4 / 5	3 / 4
45°	$\sqrt{2}/2$	$\sqrt{2}/2$	1
53°	4 / 5	3 / 5	4 / 3
60°	$\sqrt{3}/2$	1 / 2	$\sqrt{3}$
90°	1	0	∞

Newtonian Mechanics

a = acceleration	F = force
f = frequency	h = height
J = impulse	K = kinetic energy
k = spring constant	l = length
m = mass	N = normal force
P = power	p = momentum
r = radius or distance	s = displacement
T = period	t = time
U = potential energy	v = velocity or speed
W = work	x = position

Physics C

PRACTICE TEST I

SECTION I—MECHANICS

Directions: Each question listed below has five possible choices. Select the best answer given the information in each problem, and mark the corresponding oval on the answer sheet. (You may assume $g = 10$ m/s^2).

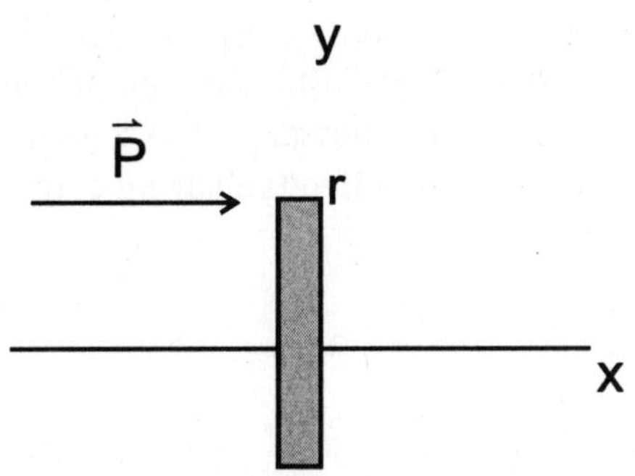

1. A uniform metal bar of length $2r$ and mass m rests on the y-axis with its center of mass at the origin, as shown above. A small projectile with momentum $\vec{p} = p_0\hat{x}$ and negligible mass strikes the bar at $y = r$ and embeds itself in the bar. What is the angular momentum of the bar after the collision?

   (A) $mrp_0\hat{z}$

   (B) $-rp_0\hat{z}$

   (C) $rp_0\hat{z}$

   (D) $2rp_0\hat{z}$

   (E) $-2rp_0\hat{z}$

2. A canon launches a metal ball horizontally using a spring as the firing mechanism. The canon has mass m_{canon} and the ball has mass m_{ball}. The spring constant is k. Initially, the spring is compressed a distance d from its equilibrium length. What is $v_{\text{canon}}/v_{\text{ball}}$, the ratio of the canon's recoil velocity to the ball's velocity?

   (A) $\dfrac{kd^2}{m_{canon}m_{ball}}$

   (B) $-\dfrac{m_{ball}^2}{m_{canon}^2}$

   (C) $\dfrac{1}{2}$

   (D) 1

   (E) $-\dfrac{m_{ball}}{m_{canon}}$

3. Consider a simple pendulum consisting of a mass m suspended by a light string of length L with a natural frequency f. If the pendulum's string is doubled in length, what is the new frequency?

(A) $f' = \frac{1}{\sqrt{2}} f$

(B) $f' = \sqrt{2} f$

(C) $f' = \frac{1}{2} f$

(D) $f' = 2f$

(E) $f' = f$

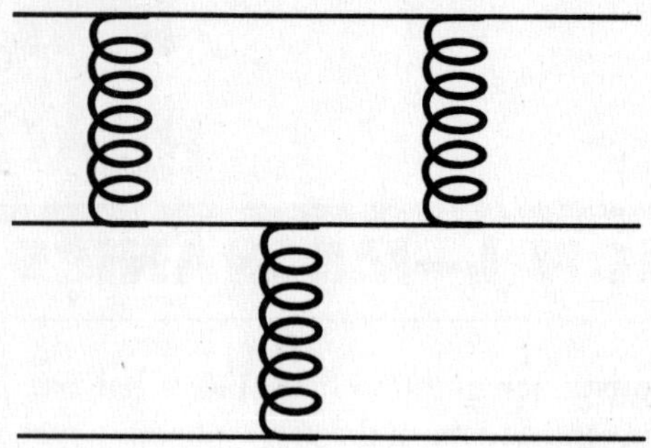

4. Each of the above three springs are identical (they have the same equilibrium length and spring constant k). They are fixed together as shown. What is the effective spring constant of the assembly?

(A) k

(B) $\frac{2}{3}k$

(C) $\frac{1}{2}k$

(D) $\frac{1}{3}k$

(E) $\frac{3}{2}k$

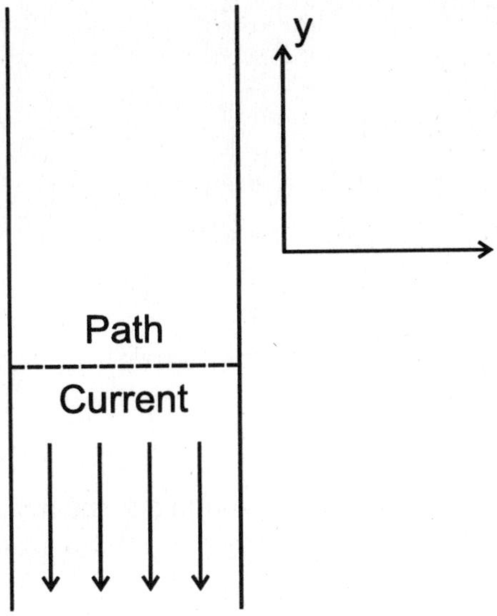

5. A person plans to swim across a river with a current of $-\frac{1}{2}\hat{y}$ m/s. In still water, the person can swim at a speed 1 m/s. What velocity must the person swim in order to traverse the river on the path shown above which is parallel to the x-axis?

(A) $\vec{v} = \hat{x}$

(B) $\vec{v} = \hat{x} + \frac{1}{\sqrt{2}}\hat{y}$

(C) $\vec{v} = \frac{\sqrt{3}}{2}\hat{x} + \frac{1}{2}\hat{y}$

(D) $\vec{v} = \hat{x} + \hat{y}$

(E) $\vec{v} = \frac{3}{2}\hat{x} + \frac{1}{2}\hat{y}$

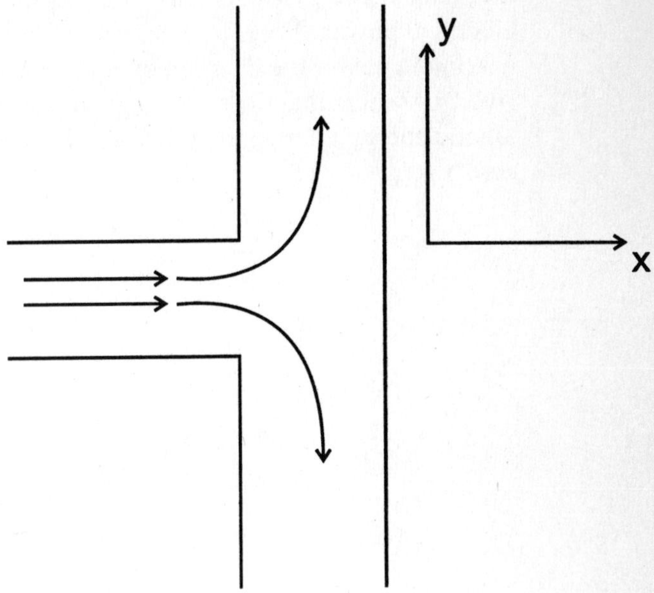

6. Consider a pipe of cross-sectional area A that meets another pipe to form a "T" as shown in the diagram above. Water flows through the pipe at a constant speed $\vec{v} = v_0\hat{x}$ and is redirected at the "T," so that its final velocity is in the y-direction. Determine the magnitude of the force on the "T" due to the water.

(A) $\rho A^2 v$

(B) $\rho A v^2$

(C) zero

(D) $\rho^2 A v^2$

(E) $\rho^2 A^2 v$

7. Water from a river flows into a dam of height h at a rate Q kg/sec. This particular dam converts mechanical energy from the river water into electrical power. What is the absolute maximum power the dam can extract from the river? (Assume that the net change in water's kinetic energy after passing through the dam is zero.)

(A) $\dfrac{Q^2 g}{h}$

(B) $\dfrac{1}{2}Qgh$

(C) $\sqrt{2Qgh}$

(D) Zero

(E) Qgh

8. A point particle moves with initial momentum $\vec{p}_i = 5\hat{x}$ (in units of kg m/s). What impulse is required to give the particle a final momentum $\vec{p}_f = 3\hat{x} + 6\hat{y}$?

(A) $-2\hat{x} + 6\hat{y}$

(B) $2\hat{x} + 6\hat{y}$

(C) $\dfrac{5}{3}$

(D) $2\hat{x} - 4\hat{y}$

(E) $-4\hat{x} + 6\hat{y}$

9. Which of the following are true ONLY of Conservative Forces?

I. $F = mA$

II. The force can be written as $\vec{F} = -\nabla U(\vec{x})$, where U is the potential energy function.

III. $U + \frac{1}{2}mv^2 = const.$ for a particle of mass m that is acted

upon by the force (assuming no external forces)

(A) Only I is true.

(B) Only II is true.

(C) Only I and II are true.

(D) Only II and III are true.

(E) I, II, and III are true.

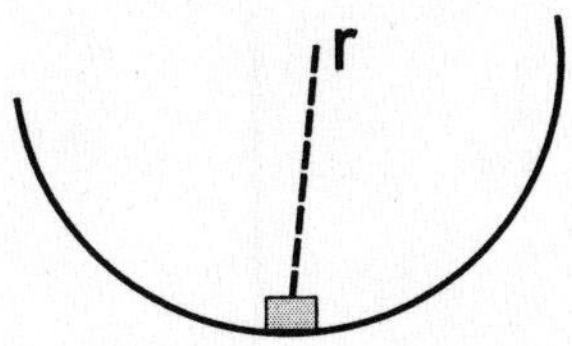

10. Consider a block of mass m that oscillates at the bottom of a bowl under the force of gravity. The bottom of the bowl has radius of curvature r as shown in the diagram above. What is the oscillation frequency (in units of radians per second)?

(A) $\dfrac{g}{r}$

(B) $\dfrac{r}{2\pi g}$

(C) $\sqrt{\dfrac{g}{r}}$

(D) $\sqrt{\dfrac{r}{g}}$

(E) $\dfrac{1}{2\pi}\sqrt{\dfrac{r}{g}}$

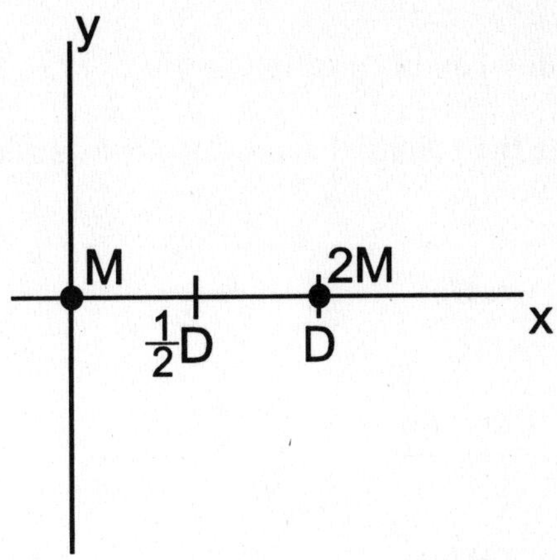

11. An astronaut of mass m floats out in space between a planet of mass M located at the origin and a second planet of mass $2M$ located at $x = D$. Determine the point at which the net gravitational force on the astronaut is zero.

(A) There is no solution.

(B) $x = \left(\sqrt{2} - 1\right)D$

(C) $x = \left(\dfrac{1}{2} + \dfrac{1}{\sqrt{2}}\right)D$

(D) $x = \dfrac{1}{\sqrt{2}}D$

(E) $x = \dfrac{1}{2}D$

12. Calculate the potential energy of the above astronaut when his position is $x = \frac{1}{2}D$. (Assume the potential energy at $x = \infty$ is zero.)

(A) $\frac{-GMm}{D^2}$

(B) $\frac{-4GMm}{D^2}$

(C) $\frac{-GMm}{6D^2}$

(D) $\frac{-6GMm}{D}$

(E) $\frac{-GMm}{4D^2}$

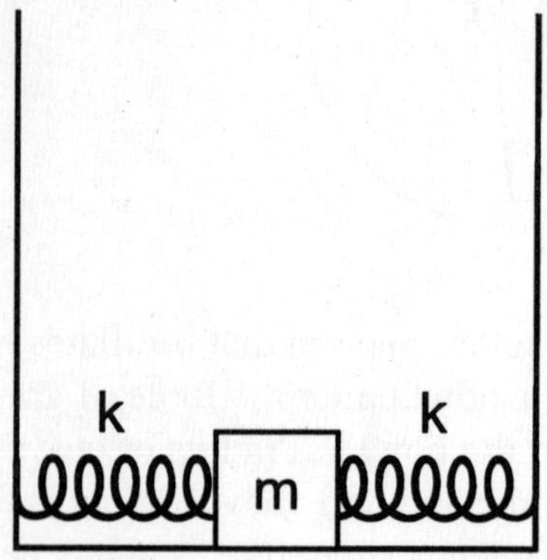

13. In the diagram above, a block of mass m sits on a frictionless surface between two fixed walls. The block is attached to each wall with identical springs of spring constant k. What is the frequency of oscillation for the above system (in units of radians per second)?

(A) $\sqrt{\frac{2k}{m}}$

(B) $\frac{1}{2\pi}\sqrt{\frac{2k}{m}}$

(C) $\sqrt{\frac{k}{m}}$

(D) $\frac{1}{2\pi}\sqrt{\frac{k}{2m}}$

(E) $\frac{1}{2\pi}\sqrt{\frac{k}{m}}$

14. A particle moves in one dimension. It's velocity is given by $v(t) = c_2t^2 + c_1t + c_{0,}$ where c_1 and c_2 are constants. What is the acceleration of the particle at time $t = 1$?

(A) $c_1 + 2c_2$
(B) zero
(C) $c_1 + c_2$
(D) c_1
(E) c_2

15. What are the units of the constant c_1 in the above equation?

(A) $\frac{\text{length}}{\text{time}}$

(B) $\frac{\text{length}}{\text{time}^3}$

(C) length

(D) length · time

(E) $\frac{\text{length}}{\text{time}^2}$

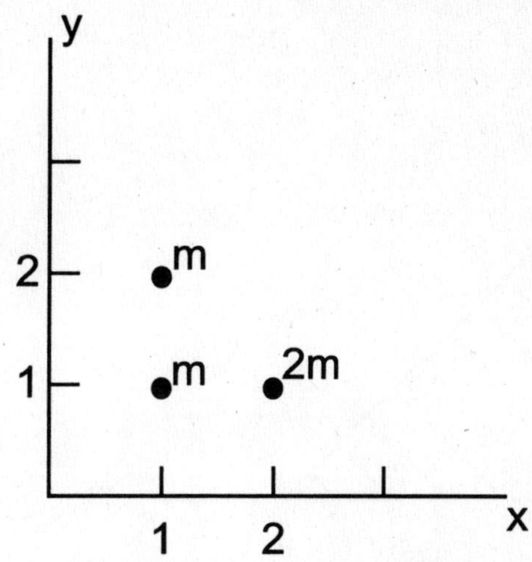

16. Solve for the center of mass of the above assembly of point particles.

(A) $6\hat{x} + 5\hat{y}$

(B) $\frac{3}{2}\hat{x} + \frac{5}{4}\hat{y}$

(C) $\frac{1}{6}\hat{x} + \frac{1}{4}\hat{y}$

(D) $2\hat{x} + \hat{y}$

(E) $\frac{1}{2}\hat{x} + \frac{1}{4}\hat{y}$

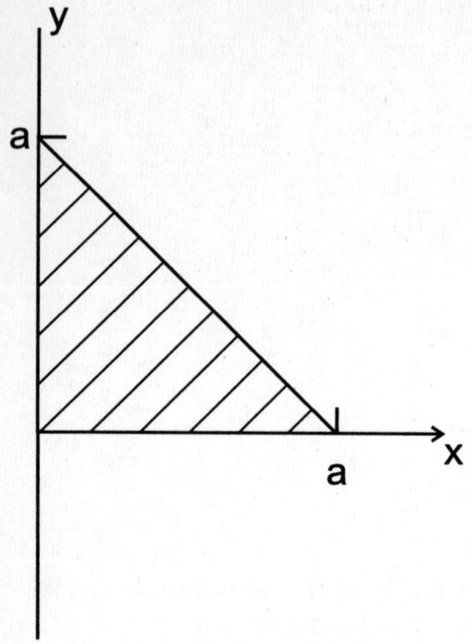

17. What is the moment of inertia for the "right" triangular sheet shown above given that the triangle has mass per unit area σ, side length a, and the moment of inertia is taken about the y-axis?

(A) $\dfrac{a^4\sigma}{12}$

(B) $\dfrac{\sigma}{6}$

(C) $a^2\sigma$

(D) $2a^4\sigma$

(E) $\dfrac{a^2\sigma}{6}$

18. The acceleration of gravity at the Earth's surface (at a radius R_e) is g. Determine the radius at which point the acceleration of gravity g' is equal to $\frac{1}{2}g$. The result should be in terms of R_e.

(A) $\dfrac{\sqrt{2}}{3}R_e$

(B) $\dfrac{3}{2}R_e$

(C) $2R_e$

(D) $\sqrt{2}R_e$

(E) Cannot be determined

19. A block of mass m moving with velocity v_i collides with a second block of mass M and that is initially at rest. Both blocks are on a frictionless surface. The blocks stick together after the collision (the collision is inelastic). In terms of m, M and v_i, determine the ratio $\dfrac{\Delta E}{E_i}$ where ΔE is the change in kinetic energy and E_i is the initial kinetic energy.

(A) $\dfrac{M}{m}$

(B) $\dfrac{M}{M+m}$

(C) $\dfrac{Mm}{M+m}$

(D) $\dfrac{2Mv_i}{M+m}$

(E) $\dfrac{M}{M-m}$

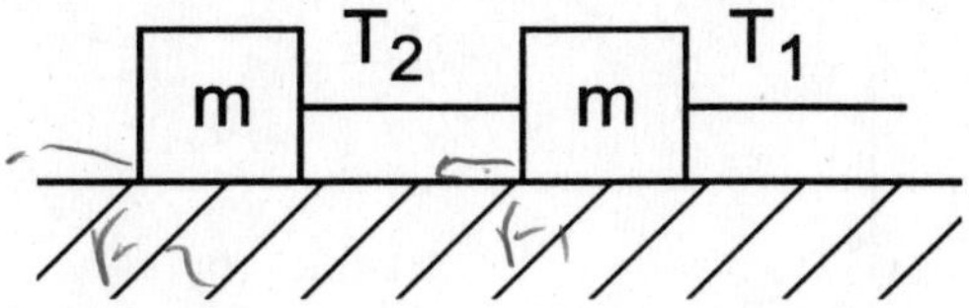

20. Two blocks rest on a flat surface. Both have a coefficient of static friction μ. The blocks are connected together by a string as shown above. Determine the minimal tension T_1 required to get both blocks moving.

(A) $2\mu g$
(B) $2\mu mg$
(C) $\frac{\mu mg}{2}$
(D) μmg
(E) Zero (Blocks move under an infinitesimal force.)

21. A small object that is initially at rest explodes into multiple particles. An observer is able to measure the momentum of three of the particles:

$\vec{p}_1 = 1\hat{x}$,

$\vec{p}_2 = 2\hat{x} + 1\hat{y}$ (in units of kg meters/second)

$\vec{p}_3 = 1\hat{y} + 2\hat{z}$

She concludes by the lack of momentum conservation that there were other particles ejected that she did not measure. Assuming that there was only one such "missing" particle, calculate the missing momentum $\vec{P}$.

(A) $(3\hat{x} + \hat{y} + 5\hat{z})$ kg m/s
(B) Zero
(C) $-(\hat{x} + 2\hat{y})$ kg m/s
(D) $-(\hat{x} + 2\hat{z})$ kg m/s
(E) $-(3\hat{x} + 2\hat{y} + 2\hat{z})$ kg m/s

22. A particle moves in one dimension. Its initial position, velocity, and acceleration are $x_0 = 2$, $v_0 = 1$, and $a_0 = 1$ (with units of m, m/s, and m/s^2). Determine the location of the particle at a time $t = 1$s.

(A) Cannot be determined
(B) $x = \frac{1}{2}$ meter
(C) $x = 4$ meters
(D) $x = 8$ meters
(E) $x = 2$ meters

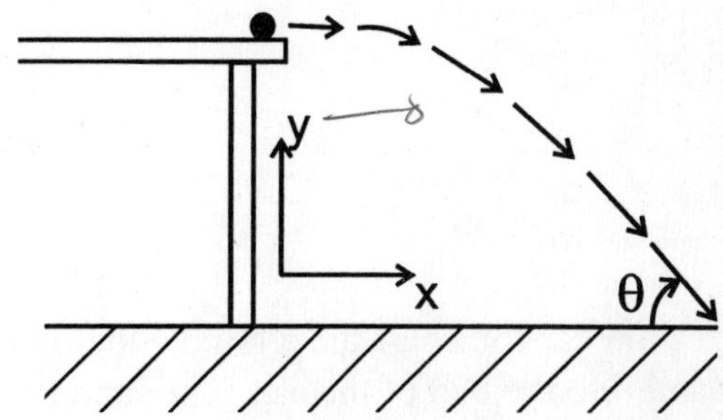

23. A ball rolls over the edge of a table with velocity $\vec{v} = v_x\hat{x}$. The surface of the table is at height h above the surface of the floor. The ball hits the floor at an angle θ to the horizontal. Determine v_x.

(A) $\sqrt{2gh}\cot(\theta)$
(B) $\sqrt{2gh}\tan(\theta)$
(C) $\sqrt{gh}$
(D) mgh
(E) $mgh\tan(\theta)$

24. A person of mass m stands with his back against a large cement block of mass M. Both the person and the block are on a frictionless floor. If a force F is applied to the cement block, causing it to move, what is the resulting force f that the block applies to the person?

(A) $\dfrac{2m}{(m-M)}F$

(B) $\dfrac{m}{(m+M)}F$

(C) $\dfrac{m}{M}F$

(D) $\dfrac{2m}{M}F$

(E) $\dfrac{m}{(m-M)}F$

25. A train car rolls beneath a hopper that pours sand into the bed of the car. The sand flows at a rate Q kg/second for a time t (assume the kinetic energy of the sand is zero). The initial velocity of the train car is v_i. What is the final velocity of the car in terms of v_i?

(A) $\left(\dfrac{m}{Qt}\right)v_i$

(B) $\left(\dfrac{1}{2}\right)v_i$

(C) $\left(\dfrac{m}{2}\right)v_i$

(D) $\left(\dfrac{m}{m+Qt}\right)v_i$

(E) $\left(\dfrac{m}{Qt}\right)^2 v_i$

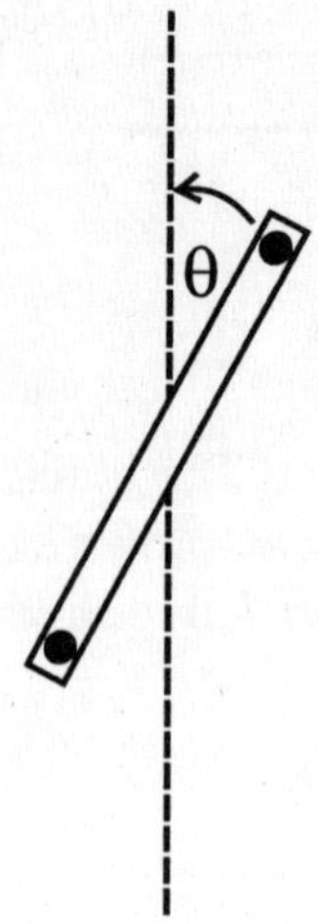

26. A uniform metal rod of mass m is fixed to a wall using two nails at each end of the rod as shown in the diagram above. The rod makes an angle θ with the vertical. Which of the following expressions most closely describes the forces on the rod from the upper and lower nails? (assume $a,b,c,d > 0$)

(A) $f_{lower} = b\hat{y},\quad f_{upper} = b\hat{y}$

(B) $f_{lower} = a\hat{x} + b\hat{y},\quad f_{upper} = -c\hat{x} + d\hat{y}$

(C) $f_{lower} = b\hat{y},\quad f_{upper} = d\hat{y}$

(D) $f_{lower} = a\hat{x} + b\hat{y},\quad f_{upper} = c\hat{x} + d\hat{y}$

(E) $f_{lower} = -a\hat{x} + b\hat{y},\quad f_{upper} = c\hat{x} + d\hat{y}$

27. The upper nail is removed, and the rod is allowed to swing. What is the initial angular acceleration of the rod about the lower nail?

(A) $\frac{g}{L}\cos(\theta)$

(B) $\frac{g}{L}\tan(\theta)$

(C) $\frac{g}{L}\cot(\theta)$

(D) $\frac{g}{2}$

(E) $\frac{3}{2}\frac{g}{L}\sin(\theta)$

28. A particle that moves in one dimension is acted upon by a force $F(x) = F_0\left(\frac{x^2}{d^2} + 2\frac{x}{d}\right)$. What is the work done to move the particle from $x = 0$ to $x = d$?

(A) $\frac{4}{3}F_0 d$

(B) $\frac{1}{3}F_0^2 d$

(C) $\frac{2}{3}F_0 d$

(D) $\frac{4}{3}F_0 d^2$

(E) $F_0 d$

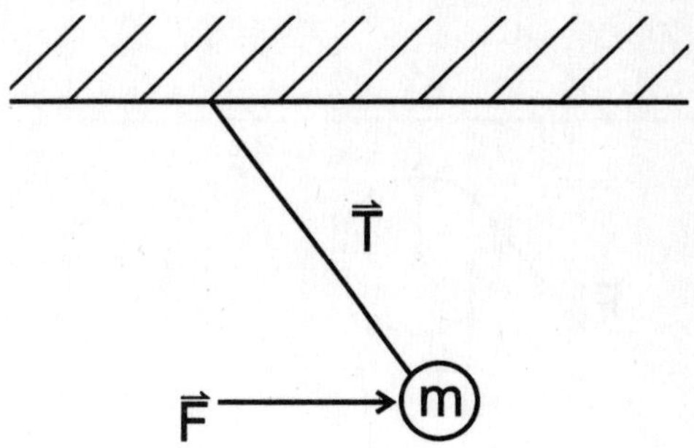

29. A sphere of mass m is suspended by a string. A force $\vec{F} = F_0\hat{x}$ is applied to the sphere as shown above. The sphere is in static equilibrium. Determine the magnitude of the tension $\vec{T}$ in the string.

(A) mg

(B) $\sqrt{(mg)^2 + 4F_0^2}$

(C) $\sqrt{mg + F_0}$

(D) $\sqrt{(mg)^2 + F_0^2}$

(E) $(mg)^2 + F_0^2$

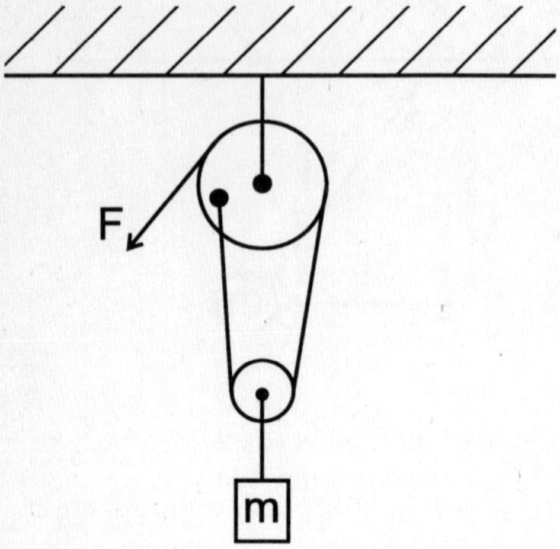

30. Consider the machine shown that involves a rope wrapped around two frictionless pulleys. Calculate the Force F required to lift an object of mass M with this machine.

(A) Mg
(B) $2Mg$
(C) $\sqrt{Mg}$
(D) $\sqrt{2Mg}$
(E) $\frac{Mg}{2}$

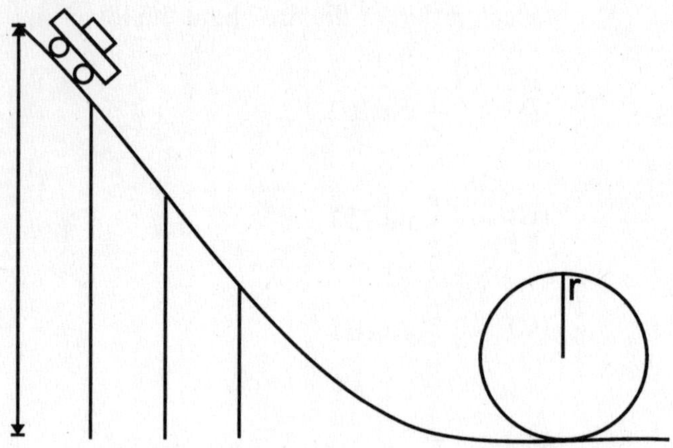

31. A toy car that rides on a frictionless track is released at a height h and rolls down the track towards a "loop de loop" of radius r. What is the minimal value of h required for the car to travel the whole loop without losing contact with the track?

(A) $2r$
(B) $\frac{7}{3}r$
(C) $\frac{5}{2}r$
(D) $\frac{7}{2}r$
(E) $\frac{1}{2}r$

32. A person of mass *m* rides in an elevator that moves upward at a constant speed 3 meters/second. The person is standing on a scale that measures the person's weight (in Newtons, of course). What is the reading on the scale?

(A) $1.3mg$
(B) mg
(C) $0.7mg$
(D) Zero
(E) Cannot be determined

33. Now suppose the elevator in the previous problem accelerates upward at a rate 0.3g. What is the reading on the scale.?

(A) $1.3mg$
(B) mg
(C) $0.7mg$
(D) Zero
(E) Cannot be determined

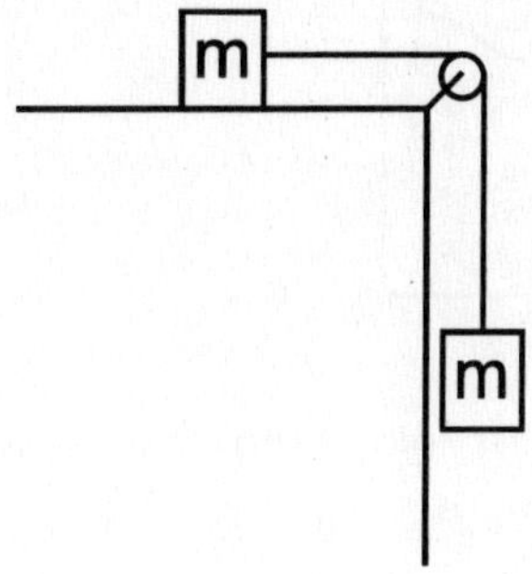

34. Two blocks, each of mass *m*, are connected by a string. The first block descends down the side of a table, and the second block slides on the table, as shown above. The coefficient of kinetic friction between the second block and the table is μ. Determine the acceleration of the two blocks.

(A) $\frac{(1-\mu)g}{2}$
(B) $(2-\mu)g$
(C) μmg
(D) $2\mu mg$
(E) $\frac{g}{2\mu}$

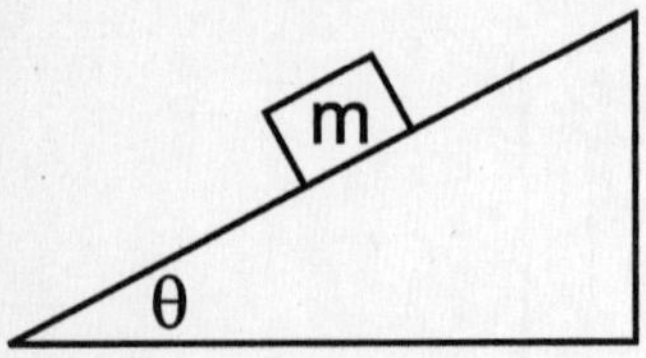

35. A block of mass m rests on a ramp that makes an angle θ with the horizontal. The coefficient of static friction between the ramp and block is μ. What is the minimal angle required for the block to slide down the ramp?

(A) $\theta = \arcsin(\mu)$

(B) $\theta = \arccos(\mu)$

(C) $\theta = \frac{\pi}{4}$

(D) $\theta = \frac{\pi}{6}$

(E) $\theta = \arctan(\mu)$

SECTION I—ELECTRICITY AND MAGNETISM

36. A positive point charge $+q$ is placed at the origin. There is an electric field

$E(x) = E_0\left(2\frac{x}{d} + 3\frac{x^2}{d^2}\right)$ that accelerates the

point charge along the x-axis. Determine the energy of the charge when it reaches the position $x = 2d$.

(A) $6qdE_0$

(B) $12q$

(C) $12qdE_0$

(D) $24qdE_0$

(E) $6qE_0$

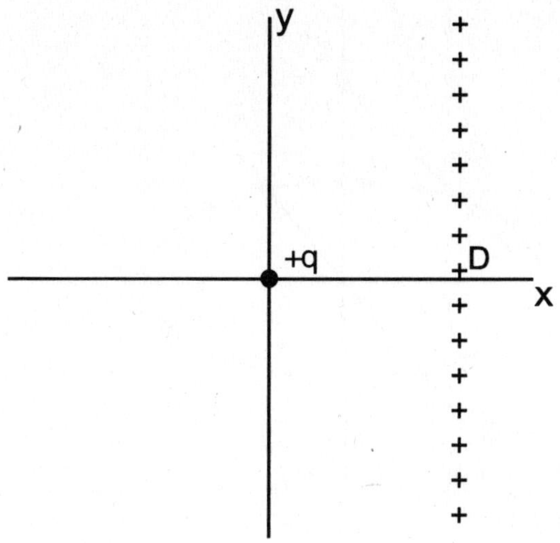

37. There is a plane of uniform positive charge density σ parallel to the yz plane and located at $x = 2d$. A point charge $+q$ is placed at the origin. Solve for the position x along the x-axis where a positive test charge will have a net force of zero.

(A) $x\dfrac{2\pi\sigma}{d}$

(B) No solution

(C) $x = \sqrt{\dfrac{q}{2\pi\sigma}}$

(D) $x = -2d$

(E) $x = d$

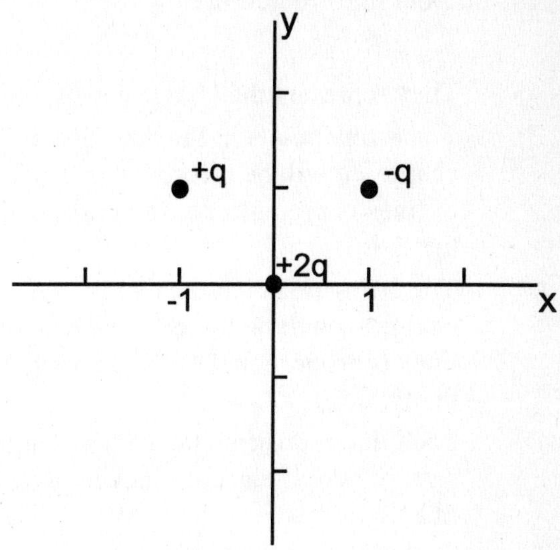

38. Three point charges are placed in the x-y plane:

$+q$ at $(-d,d)$

$-q$ at (d,d)

$+2q$ at $(0,0)$

Determine the potential energy of this configuration (assume that empty space has a potential energy of zero).

(A) $\dfrac{1}{4\pi\varepsilon_0}\dfrac{-4q^2}{d}$

(B) $\dfrac{1}{4\pi\varepsilon_0}\dfrac{2q^2}{d}$

(C) $\dfrac{\mu_0}{4\pi}\dfrac{q^2}{d}$

(D) $\dfrac{\mu_0}{4\pi}\dfrac{-4q^2}{d}$

(E) $\dfrac{1}{4\pi\varepsilon_0}\dfrac{-q^2}{2d}$

39. Which of the following best describes what electric field lines represent?

(A) They represent the charge density.

(B) The arrow points in the direction a charge $+q$ will be forced. The density of lines is proportional to the strength of the field.

(C) The arrow points in the direction a charge $-q$ will be forced. The density of lines is proportional to the strength of the field.

(D) Each line represents the path a charge $+q$ will take if placed somewhere on the line.

(E) Each line represents the path a charge $-q$ will take if placed somewhere on the line.

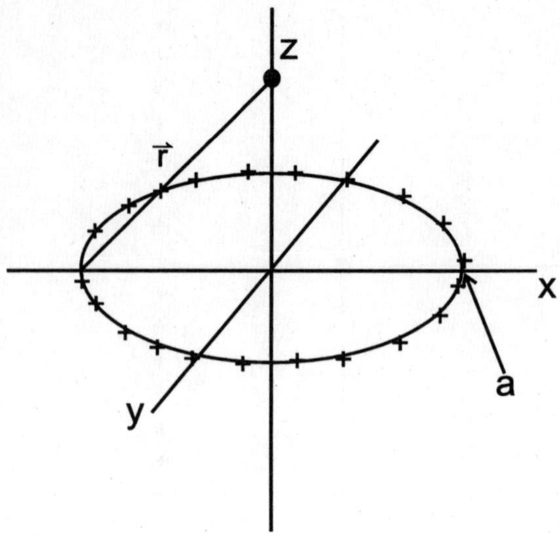

40. Solve for the electric potential on the axis of a circular ring of charge Q and radius a.

(A) $V(z)=\dfrac{Q}{4\pi\varepsilon_0}\dfrac{1}{\left(a^2+z^2\right)^{\frac{3}{2}}}$

(B) $V(z)=\dfrac{Q}{4\pi\varepsilon_0}\dfrac{1}{a(a-z)}$

(C) $V(z)=\dfrac{Q}{4\pi\varepsilon_0}\dfrac{1}{\sqrt{a^2+z^2}}$

(D) $V(z)=\dfrac{Q}{4\pi\varepsilon_0}\dfrac{1}{z(a+z)}$

(E) $V(z)=\dfrac{Q}{4\pi\varepsilon_0}\dfrac{1}{z^2}$

41. An imaginary cube of side length a encloses three charges $+q$, $-2q$, and $+q$. Determine the electric flux through the surface of the cube.

(A) $2q$
(B) $4q$
(C) Zero
(D) $\frac{q}{4\pi\varepsilon_0}$
(E) $\frac{-4q}{4\pi\varepsilon_0}$

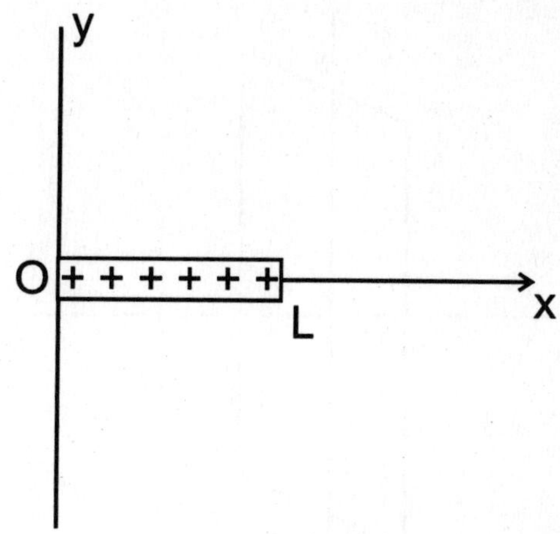

42. Charge Q is distributed uniformly over a thin rod of length L. The rod lies on the x-axis with one end at the origin and the other end at $x = L$. What is the electric field on the x-axis for $x > L$?

(A) $\frac{Q}{4\pi\varepsilon_0}\frac{1}{\left(L^2+x^2\right)^{\frac{3}{2}}}$
(B) $\frac{Q}{4\pi\varepsilon_0}\frac{1}{x(x-L)}$
(C) $\frac{Q}{4\pi\varepsilon_0}\frac{1}{(x+L)}$
(D) $\frac{Q}{4\pi\varepsilon_0}\frac{1}{\left(L^2+x^2\right)^{\frac{1}{2}}}$
(E) $\frac{Q}{4\pi\varepsilon_0}\frac{1}{\left(L^2+x^2\right)}$

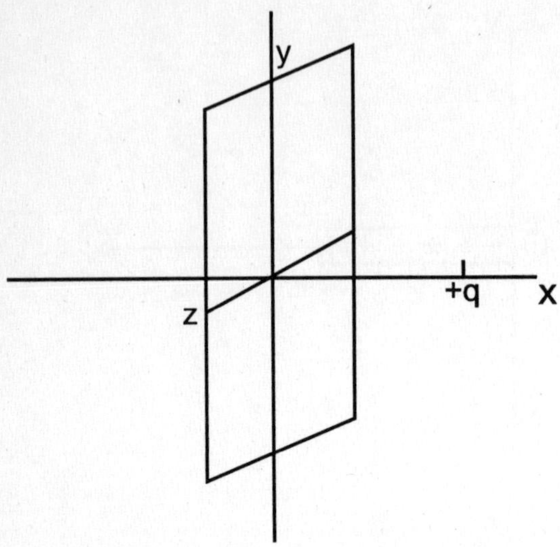

43. A plane conductor is placed in the *y-z* plane. A positive point charge $+q$ is placed on the *x*-axis at $x = x_0$. Which of the following is the electric field in the region $x > 0$?

(A) $E(\vec{r}) = \frac{-q}{4\pi\varepsilon_0}\nabla\left[\frac{1}{|\vec{r} - x_0\hat{x}|} - \frac{1}{|\vec{r} + x_0\hat{x}|}\right]$

(B) $E(\vec{r}) = \frac{-q}{4\pi\varepsilon_0}\nabla\left[\frac{1}{|\vec{r} - x_0\hat{x}|} + \frac{1}{|\vec{r} + x_0\hat{x}|}\right]$

(C) $E(\vec{r}) = \frac{-q}{4\pi\varepsilon_0}\frac{1}{|\vec{r} - x_0\hat{x}|^2}$

(D) $E(\vec{r}) = \frac{-q}{4\pi\varepsilon_0}\frac{1}{|\vec{r} - x_0\hat{x}|^2} + \frac{-2q}{4\pi\varepsilon_0}\hat{z}$

(E) $E(\vec{r}) = \frac{-q}{4\pi\varepsilon_0}\frac{1}{|\vec{r} - x_0\hat{x}|^2} + \frac{-q}{4\pi\varepsilon_0}\hat{x}$

44. Given a solid sphere of radius R centered at the origin that has charge density $\rho(\mathbf{r}) = \mathbf{c}\mathbf{r}^2$, determine the electric field everywhere.

(A) $E_{inside} = \frac{9cr^3}{5\varepsilon_0}$, $E_{outside} = \frac{9cR^5}{5\varepsilon_0 r^2}$

(B) $E_{inside} = \frac{cr^3}{5\varepsilon_0}$, $E_{outside} = \frac{cR^5}{5\varepsilon_0 r^2}$

(C) $E_{inside} = \frac{cr^3}{12\varepsilon_0}$, $E_{outside} = \frac{cR^5}{12\varepsilon_0 r^2}$

(D) $E_{inside} = \frac{1}{4\pi\varepsilon_0}r$, $E_{outside} = \frac{1}{4\pi\varepsilon_0}\frac{1}{r}$

(E) $E_{inside} = \frac{1}{4\pi\varepsilon_0}cr^2$, $E_{outside} = \frac{1}{4\pi\varepsilon_0}\frac{4c}{r}$

45. Three uniformly charged infinite planes are arranged in space as described below:

xy-plane has density $+\sigma$

yz-plane has density $+\sigma$

xz-plane has density -2σ

What is the electric field in the region $x > 0, y > 0, z > 0$?

(A) $\frac{\sigma}{2\varepsilon_0}(4\hat{z} + \hat{x} - \hat{y})$

(B) $\frac{\sigma}{2\varepsilon_0}(4\hat{z})$

(C) $\frac{\sigma}{2\varepsilon_0}(\hat{z} + \hat{x} - 2\hat{y})$

(D) $\frac{\sigma}{2\varepsilon_0}(\hat{x} - \hat{y})$

(E) $(\hat{z} + \hat{x} - \hat{y})$

46. Which of the following is true about conductors?

I. The electric field inside the conductor is always zero.

II. The electric field at the surface of the conductor is always parallel to the surface.

III. The electric field at the surface of the conductor is always perpendicular to the surface.

(A) Only I
(B) Only II
(C) Only III
(D) Only I and III
(E) Only I and II

47. Calculate the capacitance for two concentric cylinders of length L and radii a and $2a$

(A) $\dfrac{4\pi\varepsilon_0 L}{\ln 2}$

(B) $2\pi\varepsilon_0 L \sin\left(\dfrac{L}{a}\right)$

(C) $\dfrac{2\pi\varepsilon_0 L}{\ln 2}$

(D) $2\pi\varepsilon_0 L \cos\left(\dfrac{L}{a}\right)$

(E) $2\pi\varepsilon_0 L \tan\left(\dfrac{L}{a}\right)$

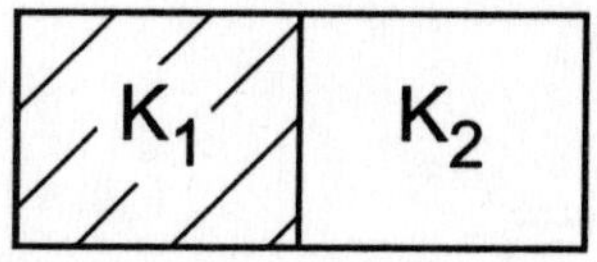

48. A parallel plate capacitor of area A and separation d is filled with two dielectric materials, both with cross-sectional areas $A/2$. The dielectric constants are k_1 and k_2. Determine the new capacitance.

(A) $(\kappa_1 + \kappa_2)\dfrac{\varepsilon_0 A}{4d}$

(B) $\kappa_1\kappa_2 \dfrac{\varepsilon_0 A}{2d}$

(C) $\dfrac{\kappa_1\kappa_2}{(\kappa_1 + \kappa_2)} \dfrac{\varepsilon_0 A}{4d}$

(D) $(\kappa_1 + \kappa_2)\dfrac{\varepsilon_0 A}{2d}$

(E) $\kappa_1\kappa_2 \dfrac{\varepsilon_0 A}{4d}$

49. Which of the following best describes the electric field just outside the surface of a conductor?

(A) Zero
(B) Parallel to the surface with magnitude $\dfrac{\sigma}{2\varepsilon_0}$
(C) Parallel to the surface with magnitude $\dfrac{\sigma}{\varepsilon_0}$
(D) Perpendicular to the surface with magnitude $\dfrac{\sigma}{2\varepsilon_0}$
(E) Perpendicular to the surface with magnitude $\dfrac{\sigma}{\varepsilon_0}$

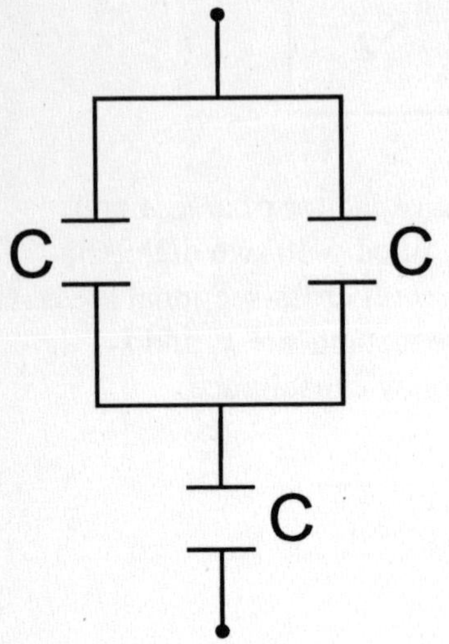

50. Solve for the equivalent capacitance of the above capacitor network.

(A) $3C$

(B) $\frac{2}{3}C$

(C) $\frac{4}{3}C$

(D) C

(E) $\frac{1}{4\pi\varepsilon_0}C$

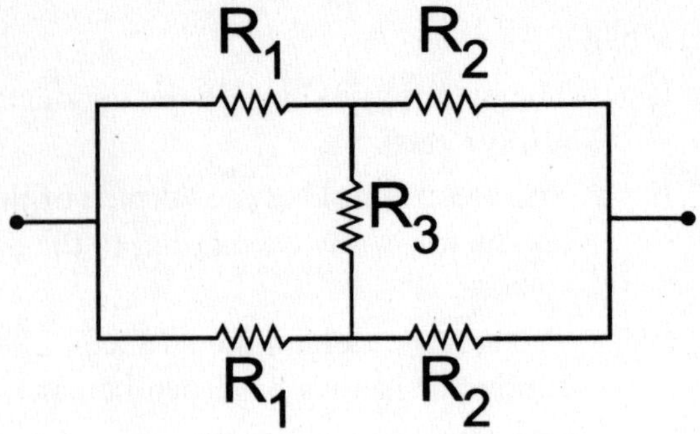

51. Solve for the equivalent resistance of the above resistor network.

(A) $R_1 + R_2$

(B) $\frac{R_1R_2}{R_1 + R_2}$

(C) $\frac{1}{3}\frac{R_1R_2}{R_1 + R_2}$

(D) $\frac{1}{2}(R_1 + R_2)$

(E) R_1R_2

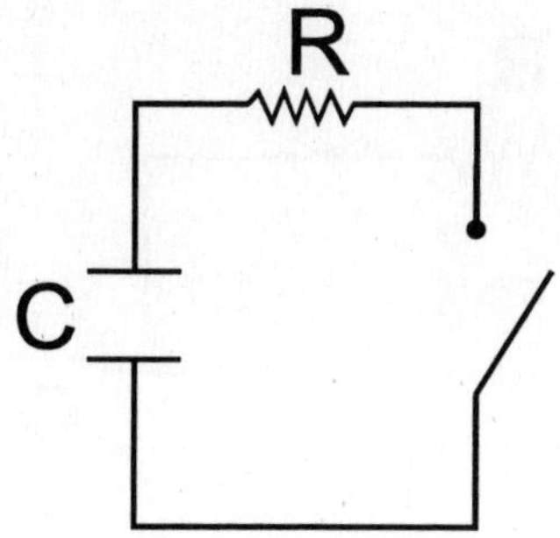

52. In the above RC circuit, a charge Q_0 is built up on the capacitor. The switch is closed and the capacitor discharges through the resistor. How long does it take the charge on the capacitor to decrease to the value $\frac{Q_0}{e^2}$ ($e \approx 2.718...$)?

(A) $2Q_0RC$

(B) $\frac{RC}{2}$

(C) $2RC$

(D) e^{2RC}

(E) $e^{RC/2}$

53. What is the energy dissipated in the above RC network when the capacitor is fully discharged?

(A) $\frac{1}{2}\frac{Q^2}{C}$

(B) CQ^2

(C) CV

(D) No energy is dissipated.

(E) Cannot be determined

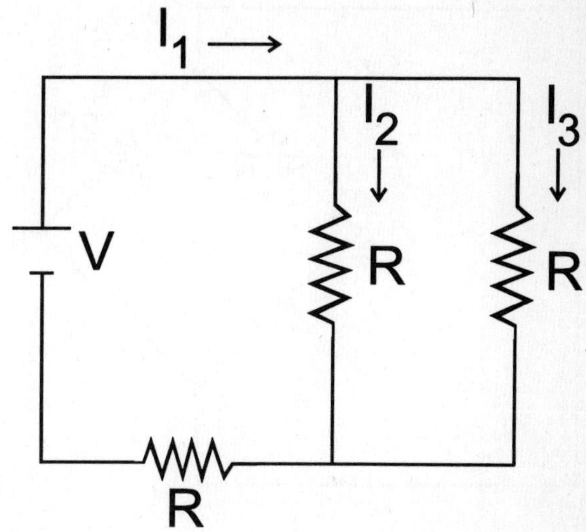

54. Determine the current everywhere in the above circuit using the directional convention shown. (The possible answers are written in the format I_1, I_2, I_3.)

(A) $\frac{4V}{3R}, \frac{2V}{3R}, \frac{2V}{3R}$

(B) $\frac{V}{3R}, \frac{V}{3R}, 0$

(C) $\frac{2V}{3R}, \frac{V}{3R}, \frac{V}{3R}$

(D) $\frac{V}{3R}, 0, \frac{V}{3R}$

(E) $\frac{V}{3R}, \frac{V}{3R}, \frac{V}{3R}$

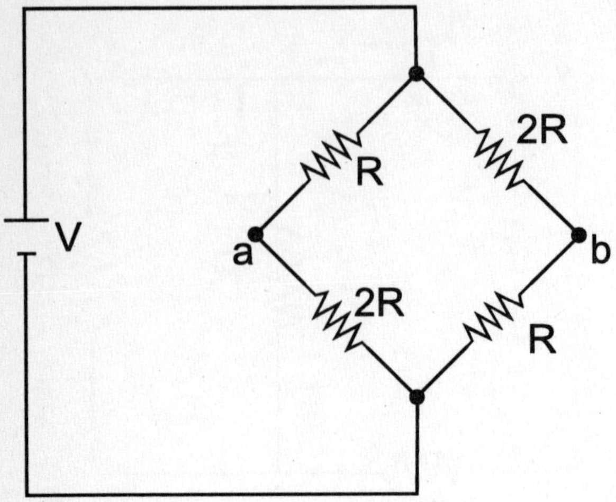

55. In the above battery-resistor network, determine $(V_a - V_b)$, the voltage between points a and b.

(A) $3V$

(B) $\frac{V}{3}$

(C) $\frac{V}{4}$

(D) $\frac{V}{2}$

(E) $\frac{3V}{4}$

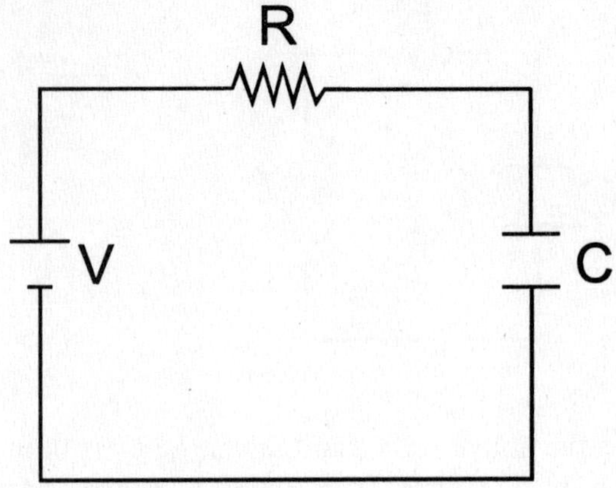

56. Which of the following represents the differential equation for the RC circuit above?

(A) $V = RC\frac{dQ}{dt}$

(B) $V = R\frac{dI}{dt} - \frac{Q}{C}$

(C) $V = R\frac{dQ}{dt} + \frac{Q}{C}$

(D) $V = R\frac{dI}{dt} + \frac{Q}{C}$

(E) $V = R\frac{dQ}{dt} - \frac{Q}{C}$

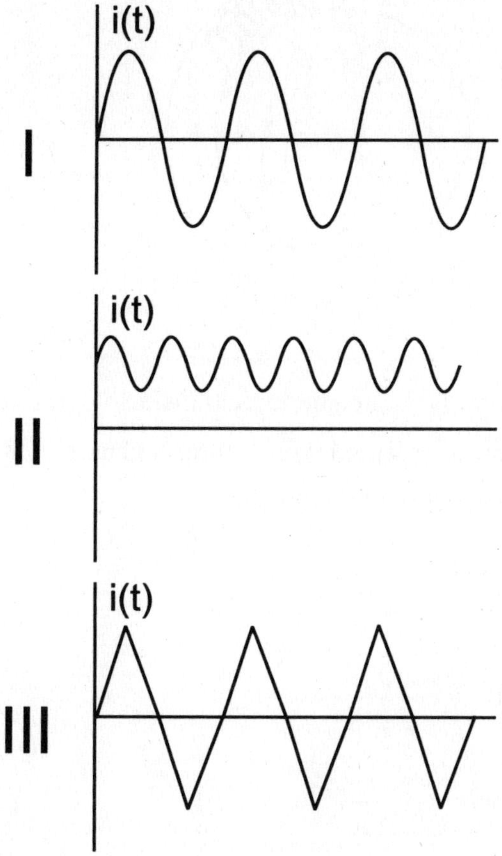

57. Which of the current signals $i(t)$ shown above would never be seen as the current across a capacitor (assume the signals are periodic)?

(A) I
(B) II
(C) III
(D) I and II
(E) I and III

58. A particle of charge q moves with velocity $\vec{v} = v_0(\hat{x} - 3\hat{z})$ in a space where there exists a constant magnetic field $\vec{B} = B_0(-2\hat{x} + \hat{y})$. What is the force on the particle?

(A) $qv_0B_0(\hat{x} + 2\hat{y})$

(B) $qv_0B_0\hat{x}$

(C) $qv_0B_0(\hat{x} + 2\hat{y} + 4\hat{z})$

(D) $qv_0B_0(\hat{z} + 6\hat{y} + 3\hat{x})$

(E) $qv_0B_0(\hat{x} - 2\hat{z})$

59. Which of the following velocities, if placed in the magnetic field of the previous problem, would result in zero force on the particle?

(A) $v_0(-\hat{x} + \frac{1}{2}\hat{y})$

(B) $v_0(\hat{x} + 2\hat{y})$

(C) $v_0(\hat{z} + 6\hat{y})$

(D) $v_0(\hat{x} - 2\hat{z})$

(E) $v_0(-2\hat{x} + \frac{1}{2}\hat{y})$

60. A wire carrying a constant current I lays flat in the xy plane in a region with a magnetic field $\vec{B} = B_0\hat{z}$. The wire can be parameterized by the coordinates $(x(\ell), y(\ell))$ where $(x(0), y(0)) = (0,0)$ corresponds to the beginning of the wire and $(x(L), y(L)) = (x_f, y_f)$ corresponds to the end of the wire. Which of the following represents the total force on the wire?

(A) $IB_0(L\hat{y} - L\hat{x})$

(B) $IB_0(-x_f\hat{y} + y_f\hat{x})$

(C) $IB_0(-L\hat{y} + L\hat{x})$

(D) $IB_0(x_f\hat{y} + y_f\hat{x})$

(E) Cannot be determined

61. Two long straight parallel wires carry currents of the same direction and magnitude I. They are separated by a constant distance d. What is the force per unit length between the wires?

(A) Zero force

(B) $\frac{\mu_0 I^2}{2\pi L}$ repulsive force

(C) $\frac{\mu_0 I^2}{2\pi L}$ attractive force

(D) $\frac{\mu_0 I^2}{2\pi d}$ repulsive force

(E) $\frac{\mu_0 I^2}{2\pi d}$ attractive force

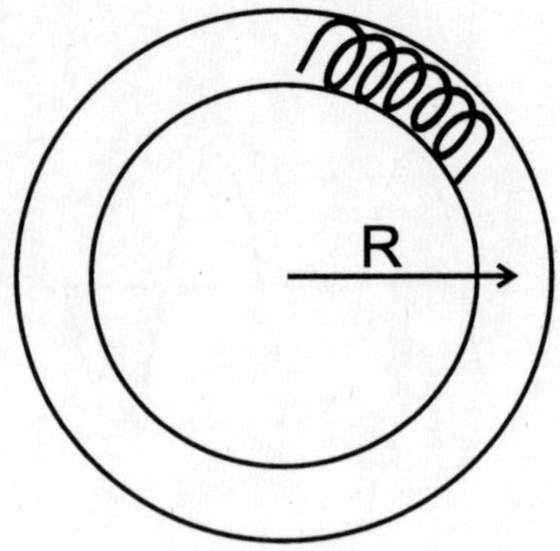

62. What is the magnetic field inside of a toroidal shaped solenoid with N turns and current I? The radius of the toroid is R.

(A) $\frac{\mu_0 I}{4\pi NR}$

(B) $\frac{\mu_0 NI}{2\pi R}$

(C) $\frac{\mu_0 I}{2\pi NR}$

(D) $\frac{3N\mu_0 I}{2\pi R}$

(E) $\frac{\mu_0 I}{2\pi R}$

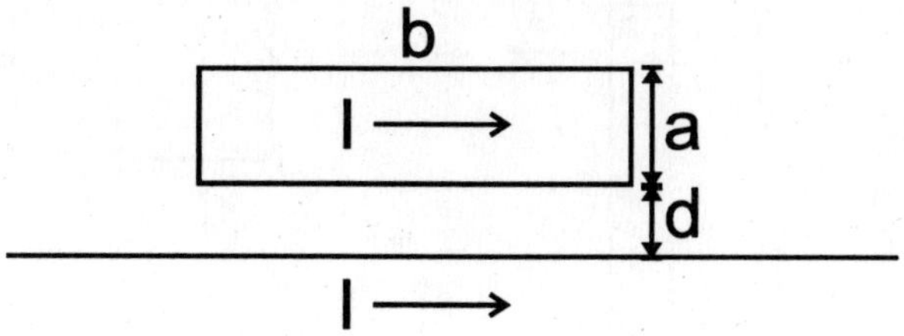

63. A current I flows down a long straight wire. Parallel to the wire is a rectangular loop of width b and height a that has the same current I and whose nearest side is a distance d away from the straight wire. Determine the total force on the loop due to the magnetic field from the wire.

(A) $\dfrac{b\mu_0 I^2}{2\pi}\left[\dfrac{1}{d+a}-\dfrac{1}{d}\right]\hat{z}$

(B) $\dfrac{b\mu_0 I^2}{2\pi}\left[\dfrac{1}{d+a}\right]\hat{y}$

(C) $\dfrac{b\mu_0 I^2}{2\pi}\left[\dfrac{1}{d-a}-\dfrac{1}{d+a}\right]\hat{z}$

(D) $\dfrac{b\mu_0 I^2}{2\pi}\dfrac{1}{d}\hat{y}$

(E) $\dfrac{b\mu_0 I^2}{2\pi}\left[\dfrac{1}{d+a}-\dfrac{1}{d}\right]\hat{y}$

64. Which of the following represents the magnitude of the magnetic field at the center (and in the plane) of a current loop of radius R?

(A) $\dfrac{\mu_0 I}{2R}$

(B) $\dfrac{\mu_0 I}{2R^2}$

(C) $\dfrac{\mu_0 I^2}{4\pi R^2}$

(D) $\dfrac{\mu_0 I^2}{2\pi R}$

(E) $\dfrac{\mu_0 I}{8\pi R}$

65. A conducting loop of area A is placed flat in the xy plane. There is a constant magnetic field $\vec{B} = B_0\hat{z}$ that passes through the loop. If the loop is set rotating about the y-axis at a constant angular velocity ω, what is the induced EMF in the loop?

(A) $V(t) = B_0 \sin(\omega t)$

(B) $V(t) = -AB_0 \cos(\omega t)$

(C) $V(t) = AB_0 \tan(\omega t)$

(D) $V(t) = \omega AB_0 \sin(\omega t)$

(E) $V(t) = AB_0\omega t$

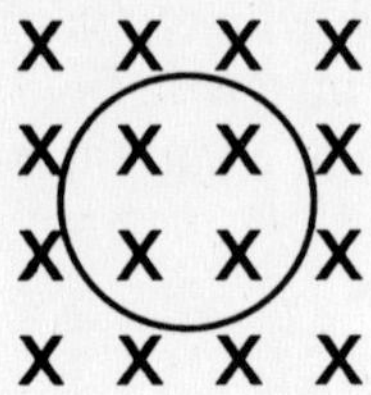

66. A conducting loop is placed in a constant magnetic field, as shown above. The magnitude of the field is suddenly decreased. Which of the following statements are true about the situation?

I. The current induced in the loop is clockwise.

II. The current induced in the loop is counterclockwise.

III. The flux from the induced current opposes the change in the external magnetic field's flux .

(A) Only I
(B) Only II
(C) Only I and III
(D) Only II and III
(E) Only III

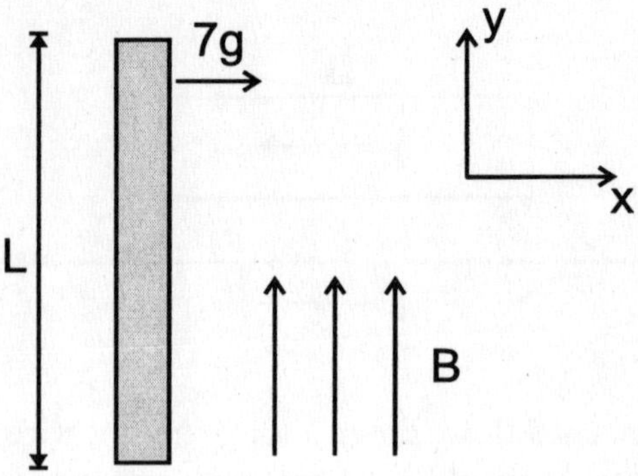

67. A conducting bar of length *L* moves with constant velocity $v_0\hat{x}$ through a constant magnetic field $-B_0\hat{z}$, as shown above. Solve for the induced electric field inside the bar.

(A) $-vB_0\hat{y}$

(B) $-vB_0\hat{x}$

(C) $vB_0\hat{x}$

(D) $2vB_0\hat{y}$

(E) $vB_0\hat{y}$

68. Which of Maxwell's equations tells us that magnetic charges do NOT exist?

I $\nabla \cdot \vec{E} = \dfrac{\rho}{\varepsilon_o}$

II. $\nabla \cdot \vec{B} = 0$

III. $\nabla \times \vec{E} = -\dfrac{d\vec{B}}{dt}$

IV. $\nabla \times \vec{B} = \mu_0 \vec{J} + \varepsilon_o \mu_0 \dfrac{d\vec{E}}{dt}$

(A) Only I
(B) Only II
(C) Only III
(D) Only IV
(E) Only III and IV

69. Which of the following is true about self inductance?

I. The self inductance of a circuit is defined to be $L = \frac{\Phi_B}{V}$ where Φ_B is the magnetic flux through the element and V is the voltage across the element.

II. The induced back EMF in the circuit due to an inductance L is $V_I = -L\frac{dI}{dt}$ where I is the current in the circuit.

III. The magnetic energy stored in a self-inductor is $U_B = \frac{1}{2}LI^2$

(A) Only I
(B) Only II
(C) Only III
(D) Only II and III
(E) Only I and II

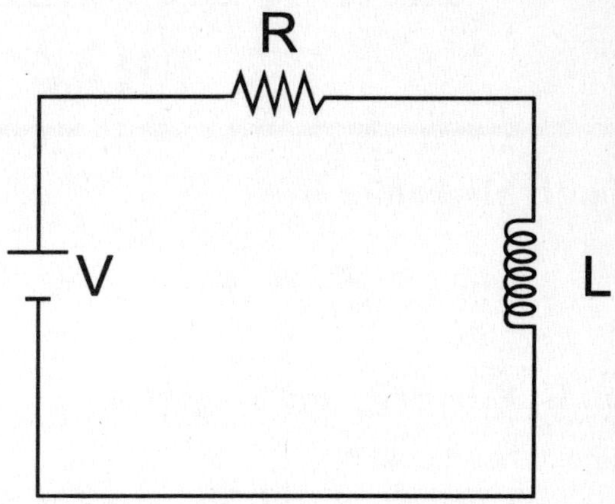

70. Which of the following represents the differential equation for the above LR circuit?

(A) $V_0 = L\frac{dI}{dt} - IR$

(B) $V_0 = L\frac{dI}{dt}$

(C) $V_0 = -L\frac{dI}{dt} - IR$

(D) $V_0 = R\frac{dI}{dt} - IL$

(E) $V_0 = L\frac{dI}{dt} + IR$

ANSWER SHEET FOR PHYSICS C PRACTICE TEST I

Section II: Mechanics

1 ⊂A⊃ ⊂B⊃ ⊂C⊃ ⊂D⊃ ⊂E⊃ 2 ⊂A⊃ ⊂B⊃ ⊂C⊃ ⊂D⊃ ⊂E⊃ 3 ⊂A⊃ ⊂B⊃ ⊂C⊃ ⊂D⊃ ⊂E⊃

Section II: Electricity and Magnetism

1 ⊂A⊃ ⊂B⊃ ⊂C⊃ ⊂D⊃ ⊂E⊃ 2 ⊂A⊃ ⊂B⊃ ⊂C⊃ ⊂D⊃ ⊂E⊃ 3 ⊂A⊃ ⊂B⊃ ⊂C⊃ ⊂D⊃ ⊂E⊃

SECTION II—MECHANICS

Directions: Answer all three questions. You will have 45 minutes in which to answer all of the questions. Note that each part within a question may not have equal weight.

Mech-1

For a ball to roll on a flat surface without slipping or skidding, the velocity of the ball's center of mass v and its angular velocity ω must satisfy the equation $v = r\omega$ (for motion on a straight path).

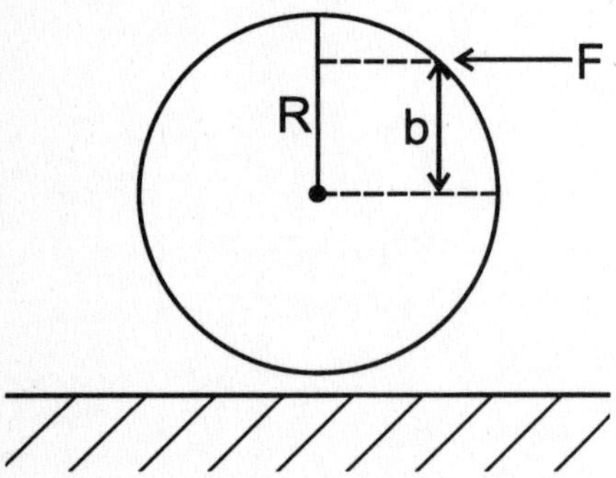

Consider a uniform, solid ball of mass M and radius R that rests on a table. If one applies a horizontal force to the ball at a point on its surface, as shown in the above figure, the ball will gain some forward momentum and it will also begin rotating. Whether the ball rotates backward or forward depends upon how far above or below center the ball is struck.

Calculate the height above (below) center at which point a horizontal force will cause the ball to immediately roll without slipping. Calculate this height in terms of the ball's radius and its mass.

Mech-2

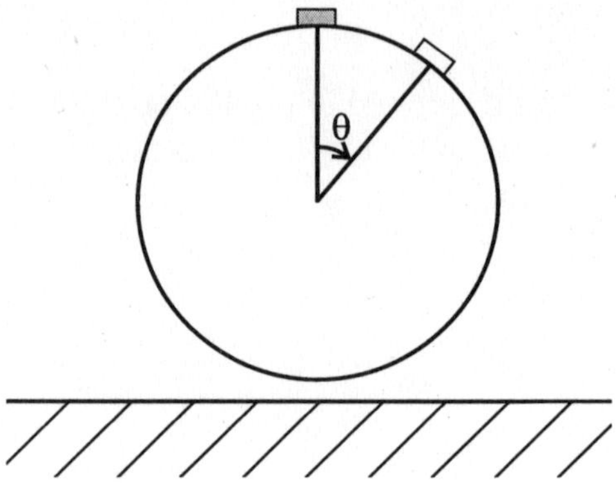

Consider a small block of mass M resting at the top of a large fixed frictionless sphere of radius R. If you perturb the block slightly, it will begin to slide down the surface of the sphere and the block will eventually lose contact with the sphere.

Calculate the exact point at which the block loses contact with the sphere in terms of M, R, and the force of gravity g. You may use any coordinate system you like (Cartesian, polar, etc.).

Mech-3

Imagine that there exists an evacuated tube running straight through the earth. If you were to let yourself fall down this tube, you would be accelerated by gravity into the center of the earth and then decelerated until you emerged at the other side of the earth with the same speed you began with. Neglecting friction, the Earth's orbit, and the Earth's rotation, you could repeat this journey over and over again in a periodic manner.

Calculate the period of oscillation (time to travel there and back). Assume the density of the earth is uniform throughout.

Calculate the maximum velocity of the person and where this occurs.

You may use:
Mearth = 5.97×10^{24}kg
Rearth = 6.38×10^{6}m

SECTION II—ELECTRICITY AND MAGNETISM

Directions: Answer all three questions. You will have 45 minutes in which to answer all of the questions. Note that each part within a question may not have equal weight.

E-1

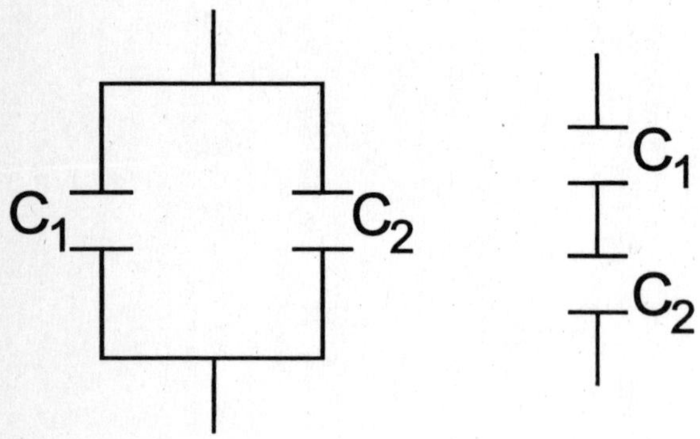

Two capacitors C_1, C_2 in parallel can be replaced by a single capacitor $C_{parallel} = C_1 + C_2$. The same two capacitors connected in series can also be replaced by a single capacitor, $C_{series} = C_1C_2/(C_1 + C)_2$. Using the fact that $Q = CV$ for all capacitors and your knowledge of circuits, derive the above combining rules.

E 2

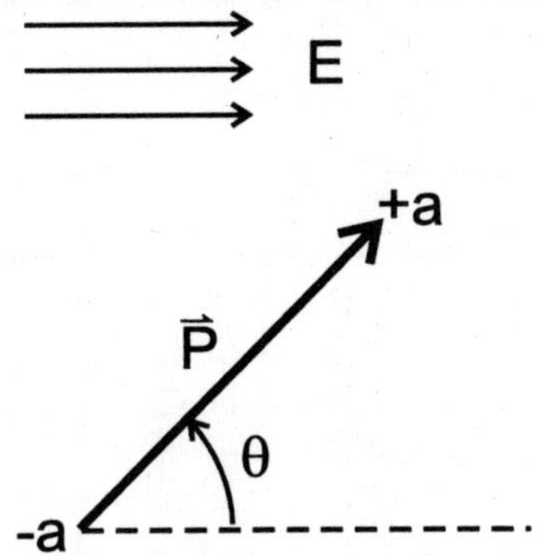

An electric dipole consists of charges $+q$ and $-q$ separated by a fixed distance a. The dipole moment vector is defined as: $\vec{p} = 2q\vec{a}$

Imagine that such a dipole is placed in an electric field $\vec{E}(x) = -\nabla V(x)$ as shown in the diagram above.

(a) Assuming for the moment that the electric field is constant $\vec{E}(x) = E_0\hat{x}$, what angular orientation of the dipole will have the lowest energy? What angular orientation will have the highest energy?

(b) Assuming that the dipole's center of mass is fixed in space, determine the potential energy of the dipole in terms of the electric field $\vec{E}(x)$ and the dipole moment vector $\vec{p}$.

(c) Is there a change in the dipole's potential energy when the dipole is translated along the *x*-axis? Under what conditions will the potential energy remain constant?

E-3

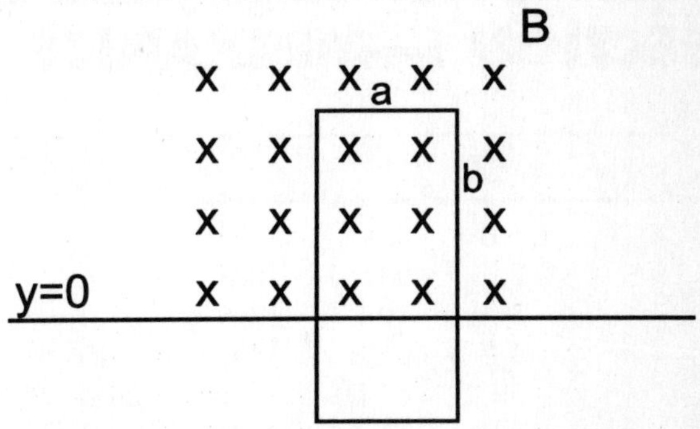

A rectangular conducting loop of width *a*, height *b*, mass *m*, and resistance *R* falls through a magnetic field:

$$\vec{B} = \begin{cases} B_0\hat{y} & \text{for } z > 0 \\ 0 & \text{for } z < 0 \end{cases}$$

Determine the terminal speed of the loop.

Physics C Practice Test 1

ANSWERS AND EXPLANATIONS

SECTION I—MECHANICS

QUICK-SCORE ANSWERS									
1.B	8.A	15.E	22.C	29.D	36.C	43.A	50.B	57.B	64.A
2.E	9.D	16.B	23.A	30.E	37.C	44.B	51.D	58.D	65.D
3.A	10.C	17.A	24.B	31.C	38.E	45.C	52.C	59.A	66.C
4.B	11.B	18.D	25.D	32.B	39.B	46.D	53.A	60.B	67.A
5.C	12.D	19.B	26.B	33.A	40.C	47.C	54.C	61.E	68.B
6.B	13.A	20.B	27.E	34.A	41.C	48.D	55.B	62.B	69.D
7.E	14.A	21.E	28.A	35.E	42.B	49.E	56.C	63.E	70.E

1. **The correct answer is (B).** Use conservation of angular momentum. The angular momentum of the rod about its center of mass is

   $\vec{L} = \vec{r} \times \vec{p} = -rp_0\hat{z}$ ·

2. **The correct answer is (E).** By conservation of momentum, $P_{canon} = -P_{ball}$,

   so $\frac{v_{canon}}{v_{ball}} = -\frac{m_{ball}}{m_{canon}}$.

3. **The correct answer is (A).** The natural frequency of a simple pendulum is

   $\sqrt{\frac{g}{L}}$, so doubling L results in a frequency $\frac{1}{\sqrt{2}}\sqrt{\frac{g}{L}}$.

4. **The correct answer is (B).** The rule for combining springs in parallel is

   $k = k_1 + k_2$. The rule for combining springs in series is $\frac{1}{k} = \frac{1}{k_1} + \frac{1}{k_2}$.

   Combine the upper two springs in parallel for an effective spring constant $k' = 2k$. Now combine this in series with the lower spring to get

   $k'' = \frac{2}{3}k$ ·

5. **The correct answer is (C).** We require the y-component of the person's velocity to cancel the y-component of the river's velocity. So

$\vec{v} = v_x\hat{x} + \frac{1}{2}\hat{y}$. To determine v_x we use $|\vec{v}| = 1$. We find $\vec{v} = \frac{\sqrt{3}}{2}\hat{x} + \frac{1}{2}\hat{y}$

6. **The correct answer is (B).** Use Newton's second law, $F = \frac{dp}{dt}$. During a time dt, a column of water of length vdt and momentum $dp = \rho vA(vdt)$ has reached the intersection. All this momentum is transferred to the "T." By simple manipulation of the previous formula, we find $\frac{dp}{dt} = \rho Av^2$.

7. **The correct answer is (E).** During time dt, an amount of water Qdt has passed through the dam. If the potential energy $dU = (Qdt)gh$ is completely delivered to the dam during its decent, then the maximum power is obviously $P_{max} = \frac{dU}{dt} = Qgh$.

8. **The correct answer is (A).** The impulse is $\Delta\vec{p} = \vec{p}_f - \vec{p}_i = -2\hat{x} + 6\hat{y}$.

9. **The correct answer is (D).** $F = ma$ is true for all types of forces, not just conservative forces.

10. **The correct answer is (C).** The geometry and the forces are the same as for a simple pendulum, so $\omega = \sqrt{\frac{g}{r}}$.

11. **The correct answer is (B).** The force on m is

$F = -\frac{GMm}{x^2} + \frac{G(2M)m}{(D-x)^2}$. We must solve

$F = 0 = \frac{GMm}{D^2}\left(-\frac{1}{a^2} + \frac{2}{(1-a)^2}\right)$ where $a = \frac{x}{D}$, a dimensionless variable. The numerator of the bracketed term reduces to the quadratic equation $a^2 + 2a - 1 = 0$, which gives $a = \sqrt{2} - 1$ or $x = (\sqrt{2} - 1)D$

12. **The correct answer is (D).** The potential energy is

$$U = -\frac{GMm}{x} - \frac{G(2M)m}{(D-x)}\Bigg|^{x=\frac{1}{2}D} \text{ so } U = \frac{-6GMm}{D}.$$

13. **The correct answer is (A).** The total force on the block is $F = -2kx$. So $m\ddot{x} = -2kx$, which is the differential equation for a simple harmonic oscillator whose angular frequency is $\omega = \sqrt{\frac{2k}{m}}$.

14. **The correct answer is (A).** The acceleration is $a = \frac{dv}{dt} = c_1 + 2c_2 t$. At time $t = 1$ $a = c_1 + 2c_2$

15. **The correct answer is (E).** The dimensions of $c_1 2$ must be velocity, so $[c_1] = \frac{length}{time^2}$, which is acceleration.

16. **The correct answer is (B).** We use the formula $R_{cm} = \frac{\sum \vec{r}_i m_i}{\sum m_i}$ to get

$$R_{cm} = \frac{m(\hat{x} + 2\hat{y}) + m(\hat{x} + \hat{y}) + 2m(2\hat{x} + \hat{y})}{4m} = \frac{3}{2}\hat{x} + \frac{5}{4}\hat{y}.$$

17. **The correct answer is (A).** We use the formula for moment of inertia $I = \int r^2 dm$ to get

$$I = \int_{x=0}^{x=a} \int_{y=0}^{y=a-x} x^2(\sigma dx dy)$$

$$= \sigma \int_{x=0}^{x=a} x^2(a-x)dx$$

$$= \frac{a^4\sigma}{12}.$$

18. **The correct answer is (D).** The acceleration of gravity at the surface of the Earth is $g = \frac{GM_e}{R_e^2}$. The acceleration of gravity at some radius αR_e is $g' = \frac{GM_e}{(\alpha R_e)^2}$. We require $\frac{g}{g'} = 2 = \alpha^2$, and so $\alpha = \sqrt{2}$.

19. **The correct answer is (B).** The energy lost is $\Delta E = \frac{p_i^2}{2m} - \frac{p_f^2}{2(m+M)}$.

Use $\boldsymbol{p}_i = \boldsymbol{p}_f$ to get $\Delta E = \frac{p_i^2}{2m}\left(1 - \frac{m}{(m+M)}\right)$. Now divide by the initial energy and simplify to get the result, $\frac{\Delta E}{E_i} = \frac{M}{M+m}$.

20. **The correct answer is (B).** The blocks will not move until $T_2 > \mu mg$, which requires that $T_1 > 2\mu mg$.

21. **The correct answer is (E).** We require $\vec{p}_1 + \vec{p}_2 + \vec{p}_3 + \vec{p}_4 = 0$. Plug in for $\vec{p}_1, \vec{p}_2, \vec{p}_3$ and solve.
$\vec{p}_4 = -(3\hat{x} + 2\hat{y} + 2\hat{z})$ kg m / s.

22. **The correct answer is (C).** Integrate twice to get $x(t) = x_0 + v_0 t + a_0 t^2$. Plug in for x_0, v_0, a_0, t to get $x = 4$ meters.

23. **The correct answer is (A).** First, solve for the time of flight: $h = \frac{1}{2}at^2$, so $t = \sqrt{\frac{2h}{g}}$. Now solve for the *y*-component of the velocity of the ball at the point of impact: $v_y = gt = \sqrt{2gh}$. Now use $\cot(\theta) = \frac{v_x}{v_y}$, so $v_x = \sqrt{2gh}\cot(\theta)$.

24. **The correct answer is (B).** The applied force is $F = (m + M)\,a$. Solve for A and plug this into the force equation for the person $f = m\left(\dfrac{F}{m+M}\right)$, so $\dfrac{f}{F} = \dfrac{m}{(m+M)}$.

25. **The correct answer is (D).** $p_i = p_f$, so $mv_i = (m+Qt)v_f$ and $v_f = \left(\dfrac{m}{m+Qt}\right)v_i$.

26. **The correct answer is (B).** This can be determined by considering the acceleration of the rod's center of mass when either nail is removed.

27. **The correct answer is (E).** We use the equation $\tau = I\alpha$ and plug in for the torque and the moment of inertia of the rod,

$$\frac{L}{2}mg\sin(\theta) = \left(\frac{1}{3}mL^2\right)\alpha \text{ to get } \alpha = \frac{3}{2}\frac{g}{L}\sin(\theta).$$

28. **The correct answer is (A).** The work done is

$$W = F_0 d\int_0^1 (x^2 + 2x)dx = \frac{4}{3}F_0 d.$$

29. **The correct answer is (D).** To balance the forces, $\vec{T} = +F_0\hat{x} + mg\hat{y}$, and so $|T| = \sqrt{(mg)^2 + F_0^{\,2}}$.

30. **The correct answer is (E).** The total upward force on the block is $2T$. The minimum force required to lift the the blocks is then $T = \dfrac{Mg}{2}$.

31. **The correct answer is (C).** The track must always be exerting a force on the car. If this is not true, the car will fall off. Apply this reasoning to the apex of the loop. Use $F_{track} + F_g = F_{centripetal}$ and let $F_{track} = 0$ (this is the critical point) to get $\frac{mv^2}{r} = mg$, where v is the velocity at the apex of the loop. Solve for v using the conservation of energy $mg(h-2r) = \frac{1}{2}mv^2$.

Plug the quantity v^2 into the previous equation and reduce to get $h = \frac{5}{2}r$.

32. **The correct answer is (B).** A constant velocity will not affect the reading on the scale, so $F = mg$.

33. **The correct answer is (A).** A net upward force of $0.3mg$ is required, so the scale will read $F = 1.3mg$

34. **The correct answer is (A).** Assume T is the tension in the string and a is the acceleration. Applying Newton's Second Law:
For the first block will yield $mg - T = ma$
For the second block will yield $T - \mu mg = ma$

Solving for a will yield $a = \frac{g(1-\mu)}{2}$.

The force on the second block will be $F = mg - \mu mg$, so $a = (1-\mu)g$.

35. **The correct answer is (E)**. The block will begin to slide when the component of gravity parallel to the plane is equal to the force of friction,

$F_{\parallel} \geq F_{friction}$, $mg\sin(\theta) \geq \mu mg\cos(\theta)$ of the critical angle is

$\tan(\theta) = \mu$ or $\theta = \arctan(\mu)$.

SECTION I—ELECTRICITY AND MAGNETISM

36. **The correct answer is (C).** The kinetic energy is

$$q[V(2d)-V(0)]=q\int_0^{2d}E(x)dx=qdE_0(x^2+x^3)|^{x=2}=12qdE_0\cdot$$

37. **The correct answer is (C).** Set the electric field to zero and solve:

$$\frac{q}{4\pi\varepsilon_0 x^2}-\frac{\sigma}{2\varepsilon_0}=0\text{, so } x=\sqrt{\frac{q}{2\pi\sigma}}\,.$$

38. **The correct answer is (E).** The potential energy has a term for each pair of charges: $U=\frac{q^2}{4\pi\varepsilon_0 d}\left(-\sqrt{2}+\sqrt{2}-\frac{1}{2}\right)=\frac{1}{4\pi\varepsilon_0}\frac{-q^2}{2d}$.

39. **The correct answer is (B).**

40. **The correct answer is (C).** $V(\vec{r})=\frac{1}{4\pi\varepsilon_0}\int\frac{dq}{|\vec{r}|}=\frac{1}{4\pi\varepsilon_0}\frac{1}{|\vec{r}|}\int dq$, where we have used the fact that $|\vec{r}|$ is constant for the integration over the ring. The potential is then:

$$V(z)=\frac{Q}{4\pi\varepsilon_0}\frac{1}{\sqrt{a^2+z^2}}\cdot$$

41. **The correct answer is (C).** Use Gauss's Law. The electric flux is zero.

42. **The correct answer is (B).** In the following, use $\lambda=\frac{Q}{L}$.

$$\begin{aligned}E(x)&=\frac{1}{4\pi\varepsilon_0}\int_0^L\frac{\lambda dx'}{(x-x')^2}\\&=\frac{1}{4\pi\varepsilon_0}\left[\frac{\lambda}{(x-x')}\right]_0^L\\&=\frac{\lambda}{4\pi\varepsilon_0}\left[\frac{1}{(x-L)}-\frac{1}{x}\right]\\&=\frac{Q}{4\pi\varepsilon_0}\frac{1}{x(x-L)}\end{aligned}$$

43. **The correct answer is (A).** We can replace the plane conductor by an image charge $-q$ at $x = -x_0$ without changing the resulting field. The electric field is then $E(\vec{r}) = -\nabla V(\vec{r}) = \frac{-q}{4\pi\varepsilon_0}\nabla\left[\frac{1}{|\vec{r}-x_0\hat{x}|} - \frac{1}{|\vec{r}+x_0\hat{x}|}\right]$

44. **The correct answer is (B).** Use Gauss's law to solve for the field inside the sphere:

$$
\begin{aligned}
E(4\pi r^2) &= \frac{1}{\varepsilon_0}Q_{inside} \\
&= \frac{1}{\varepsilon_0}\int_0^r \rho(r')(4\pi r'^2 dr') \\
&= \frac{4\pi c}{\varepsilon_0}\int_0^r r'^4 dr' \\
&= \frac{4\pi c r^5}{5\varepsilon_0}
\end{aligned}
$$

So inside the charged sphere the electric field is $E = \frac{cr^3}{5\varepsilon_0}$ and outside the field is $E = \frac{cR^5}{5\varepsilon_0 r^2}$.

45. **The correct answer is (C).** Superpose the fields for all three planes of charge: $\vec{E} = \frac{\sigma}{2\varepsilon_0}(\hat{z} + \hat{x} - 2\hat{y})$.

46. **The correct answer is (D).** The second statement is false. If a parallel component to the electric field existed at the surface, then the charges would move in response to this.

47. **The correct answer is (C).** To calculate the capacitance, place a charge Q on the inner cylinder and a charge $-Q$ on the outer cylinder, then calculate the potential difference between them.

$$V = \int_a^{2a} dr \cdot E$$

$$= \frac{Q}{2\pi\varepsilon_0 L}\int_a^{2a} dr \cdot r^{-1}$$

$$= \frac{Q}{2\pi\varepsilon_0 L}\ln(2)$$

Compare this to the capacitor equation $Q = CV$, and we find that

$$C = \frac{2\pi\varepsilon_0 L}{\ln 2}.$$

48. **The correct answer is (D).** Treat this as two capacitors of area $\frac{A}{2}$ in parallel. So $C = \kappa_1 \frac{\varepsilon_0 A}{2d} + \kappa_2 \frac{\varepsilon_0 A}{2d} = (\kappa_1 + \kappa_2)\frac{\varepsilon_0 A}{2d}$.

49. **The correct answer is (E).** The electric field must be perpendicular to the surface for reasons stated earlier, and the magnitude can be determined by application of Gauss' Law.

50. **The correct answer is (B).** Combine the parallel capacitors first, then the problem is reduced to two capacitors in series: $C_{eq} = \frac{2}{3}C$.

51. **The correct answer is (D).** By symmetry there will never be current through R_3, so it can be removed. The remaining problem is trivial:

$$R_{eq} = \frac{1}{2}(R_1 + R_2).$$

52. **The correct answer is (C).** The charge on the capacitor is given by $Q(t) = Q_0 e^{-t/RC}$, from which it is straightforward to find

$$Q(t = 2RC) = Q_0 e^{-2}$$

53. **The correct answer is (A).** All the energy stored in the capacitor $U = \frac{1}{2}\frac{Q^2}{C}$ is dissipated in the resistor.

54. **The correct answer is (C).** By symmetry $I_2 = I_3$. Solve for I_1 by first solving for the equivalent resistance of the circuit: $R_{eq} = \frac{3}{2}R$, so $I_1 = \frac{2V}{3R}$.

55. **The correct answer is (B).** By symmetry the current through each branch is the same, so $(V_a - V_b) = (V - RI) - (V - 2RI) = RI$, where I is the current through each branch. The total current satisfies $V = R_{eq}I_{total} = \left(\frac{3}{2}R\right)(2I)$. We solve for I in terms of V and plug this into the first equation to get our result, $(V_a - V_b) = RI = \frac{V}{3}$.

56. **The correct answer is (C).** Use Kirchoff's voltage loop rule:

$$V = R\frac{dQ}{dt} + \frac{Q}{C}.$$

57. **The correct answer is (B).** The second current signal would never be seen because its average value is nonzero. This would mean that the charge on the capacitor increases without band, which is impossible.

58. **The correct answer is (D).**

$$\vec{F} = q\vec{v}\times\vec{B} = qv_0B_0(\hat{x} - 3\hat{z})\times(-2\hat{x} + \hat{y})$$
$$= qv_0B_0(\hat{z} + 6\hat{y} + 3\hat{x})$$

59. **The correct answer is (A).** A velocity parallel to the magnetic field will not have a force on it, so $\vec{v} = v_0(-\hat{x} + \frac{1}{2}\hat{y})$.

60. **The correct answer is (B).** Use the magnetic force law, $d\vec{F} = \vec{I}d\ell\times\vec{B}$, and split up the differential into *x* and *y* components:

$d\vec{F} = IB_0[dx(\hat{x}\times\hat{z}) + dy(\hat{y}\times\hat{z})]$. When we add up all contributions to the force, notice that a path in the positive *x*-direction will exactly cancel an equal length path in the negative *x*-direction. Consequently, no matter how curvy the path is, it can be reduced to the following two legs:

$$\vec{F} = IB_0\left[\int_0^{x_f} dx(\hat{x}\times\hat{z}) + \int_0^{y_f} dy(\hat{y}\times\hat{z})\right] = IB_0\left(-x_f\hat{y} + y_f\hat{x}\right) \cdot$$

61. **The correct answer is (E).** The wires are attracted to each other with a force, $\left|\vec{I}L\times\vec{B}\right| = (IL)\left(\frac{\mu_0 I}{2\pi d}\right)$, so the force per unit length is $\frac{F}{L} = \frac{\mu_0 I^2}{2\pi d}$.

62. **The correct answer is (B).** Use Ampere's law with a closed loop of radius R taken around the center of the toroid. In this case $\oint \vec{B}\cdot d\vec{s} = \mu_0 I_{enc}$ reduces to $2\pi RB = \mu_0 NI$, so the magnetic field inside the toroid is $B = \frac{\mu_0 NI}{2\pi R}$.

63. **The correct answer is (E).** The contribution to the force from the sides of length a cancel each other. This leaves contributions for the other two sides of the loop. $\vec{F} = \frac{b\mu_0 I^2}{2\pi}\left[\frac{1}{d+a} - \frac{1}{d}\right]\hat{y}$, where we've used $B = \frac{\mu_0 I}{2\pi d}$ for the magnetic field from a long straight wire of current I at a distance d from the wire.

64. **The correct answer is (A).** $B = \frac{\mu_0}{4\pi}\int \frac{Id\vec{s}\times\hat{r}}{r^2} = \frac{\mu_0 I}{4\pi}\int_0^{2\pi}\frac{Rd\theta}{R^2} = \frac{\mu_0 I}{2R}$.

65. **The correct answer is (D).** The magnetic flux is given by $\Phi_B = \vec{B}\cdot\vec{A} = AB_0\cos(\omega t)$. The induced EMF in the loop is $V(t) = -\frac{d\Phi_B}{dt} = \omega AB_0\sin(\omega t)$.

66. **The correct answer is (C).** The induced current is clockwise. III is just Lenz's law, which is true.

67. **The correct answer is (A).** Positive charge carriers in the rod undergo a force $\vec{F} = qvB_0\hat{y}$ and migrate to the top of the bar until there exists an induced electric field. A steady state exists when the electric force balances the magnetic force $qE = -qvB_0$. The induced electric field is then $E = -vB_0\hat{y}$.

68. **The correct answer is (B).** This is basically the magnetic equivalent of Gauss' Law, except there is no source term allowed—hence, no magnetic charge is allowed.

69. **The correct answer is (D).** $L = \frac{\Phi_B}{V}$ is not true. The correct statement is

$$L = \frac{\Phi_B}{I}.$$

70. **The correct answer is (E).** Use Kirchoff's voltage loop rule:

$$V_0 = L\frac{dI}{dt} + IR.$$

SECTION II—MECHANICS

Mech-1

The "no-slip" condition can be written in two ways:

$V = R\omega$ and $a = R\alpha$, assuming $V_0 = R\omega_0$, where *V* and *A* are the linear velocity and acceleration and ω, α are the angular velocity and acceleration. The second equation is the time derivative of the first equation, and it is equivalent to the first if the initial velocity and angular velocity satisfy the no-slip condition.

Recognize the following facts:

$F = Ma$

$\tau = I_{sphere}\alpha$

$\tau = \vec{R} \times \vec{F} = bF$,

where τ is the torque on the ball and $I_{sphere} = \frac{2}{5}MR^2$ is the moment of inertia of the ball. We equate the right-hand sides of the last two equations to get, $Fb = I_{sphere}\alpha$.

Now, plugging in for *F* and solving for *b*, we have $b = \frac{I_{sphere}\alpha}{Ma}$.

Finally, plug in for α using the "no-slip" condition, and we get the desired result, $b = \frac{I_{sphere}}{MR} = \frac{2}{5}R$.

Mech-2

The condition for the block to remain touching the sphere is

$F \geq F_c$

$F \geq \frac{MV^2}{R}$,

where *F* is the component of the force of gravity that points toward the center of the sphere and F_c is the centripetal force required to keep the block moving in circle of radius *R*. We can solve for *F*.

$F = Mg\cos(\theta)$ and the contact condition becomes $g\cos(\theta) \geq \frac{V^2}{R}$,

where V is the velocity of the block at the angle θ. We can solve for V using conservation of energy.

$$\frac{1}{2}MV^2 = Mg(R - R\cos(\theta)).$$

Plug this into the above inequality, and reduce to get the condition

$\cos(\theta) \geq \frac{2}{3}$,

so the threshold angle is $\theta_{critical} = \arccos(\frac{2}{3})$.

Mech-3

Start with Newton's third law: $F_{gravity} = M_{person}A$. $F_{gravity}$ becomes weaker as the person gets closer and closer to the center of

the earth. More precisely, $F_{gravity} = \frac{GM_{enclosed}(r)M_{person}}{r^2}$, where

$M_{enclosed}$ is the mass of the sphere enclosed by the radius r. Assuming the earth is uniform, we have

$$M_{enclosed} = \tfrac{4}{3}\pi r^3\left(\frac{M_{earth}}{\tfrac{4}{3}\pi R_{earth}^{\ 3}}\right)$$
$$= M_{earth}\left(\frac{r^3}{R_{earth}^{\ 3}}\right)$$

so

$$F_{gravity} = \left(\frac{GM_{person}M_{earth}}{R_{earth}^{\ 3}}\right)r,$$

and plugging this into Newton's third Law gives

$$\ddot{r} + \left(\frac{GM_{earth}}{R_{earth}^{\ 3}}\right)r = 0,$$

which we recognize as the simple harmonic oscillator equation. The frequency of oscillation is then

$$\omega = \sqrt{\left(\frac{GM_{earth}}{R_{earth}^{\ 3}}\right)} \cong 0.00124 \tfrac{rad}{sec}$$

This corresponds to a period $T = \dfrac{2\pi}{\omega} \cong 5072\,\text{sec}$ or approximately 84.5 minutes. Now we calculate the maximum velocity

$$r = R_{earth}\cos(\omega t)$$
$$v = -\omega R_{earth}\sin(\omega t)$$

so

$$v_{\max} = \omega R_{earth} \cong 7.94 \tfrac{km}{s}$$

or approximately 18,000 miles/hr.

SECTION II—ELECTRICITY AND MAGNETISM

E-1

(a) In the parallel configuration, C_1 and C_2 have the same voltage across them, so

$$Q_1 = C_1 V$$
$$Q_2 = C_2 V$$

The combined charge stored by the pair is $Q = Q_1 + Q_2$, so

$$\begin{aligned} C_{parallel} V &= Q \\ &= (Q_1 + Q_2). \end{aligned}$$

Plugging in for Q_1, Q_2 and factoring out V gives $C_{parallel} = C_1 + C_2$.

(b) In the series configuration, C_2 and C_1 have the same charge collected on them. This follows from the fact that the wire connecting them is electrically neutral.

$$Q = C_1 V_1$$
$$Q = C_2 V_2$$

Kirchoff's voltage loop rule requires $V = V_1 + V_2$. Hence, the combined capacitance satisfies the equation

$$\begin{aligned} Q &= C_{series} V \\ &= C_{series} (V_1 + V_2). \end{aligned}$$

Replace V_1 and V_2 in the previous equation, factor out Q, and solve for C_{series}:

$$C_{series} = \frac{C_1 C_2}{C_1 + C_2}.$$

E-2

(a) The potential energy is minimum when $\vec{p}$ is parallel to the electric field and maximum when $\vec{p}$ is anti-parallel to the field. To see this, consider the effects of the electric field on the $+q$ and $-q$ ends of the dipole.

(b) The potential energy is $U(x) = q[V(x+2a\cos\theta) - V(x)]$.

Rewrite this as $U(x) = 2aq\cos\theta\left[\frac{V(x+2a\cos\theta)-V(x)}{2a\cos\theta}\right]$.

In the limit of small a this becomes a derivative

$U(x) = 2aq\cos\theta\frac{dV(x)}{dx}$, which is our result.

$= -\vec{p}\cdot\vec{E}(x)$

(c) The potential energy of the dipole is invariant under translation only when the electric field is constant. An electric field gradient will cause a translational force on the dipole. To see this, use $\vec{F} = -\nabla U(x)$

E-3

As the loop falls, the magnetic flux through it is changing, so a current will be induced. The external magnetic field will exert a force on this current, and when this force equals the force of gravity, the loop will have reached its terminal velocity.

More precisely the magnetic flux is $\Phi_B = B_0 a\ell$, where ℓ is the portion of the side b remaining in the magnetic field. The EMF induced in the loop is

$$V = -\frac{d\Phi_B}{dt} = -B_0 a\frac{d\ell}{dt} = -B_0 av.$$

Where v is the velocity of the loop. The current in the loop is just

$I = \frac{V}{R} = \frac{-B_0 av}{R}$ and we use the right-hand rule to determine that the direction of the current is clockwise. Now calculate the force on this current. The contribution from the vertical portions of the loop cancel each other out, and the remaining contribution from the top (side of length a)

$$\vec{F} = \vec{I}d\ell \times \vec{B}$$

yields $= IaB_0\hat{y}$

$$= \frac{B_0^2 a^2 v}{R}\hat{y}$$

where we have plugged in for I in the last step. Balance this force against the force of gravity

$\frac{B_0^2 a^2 v}{R}\hat{y} - mg\hat{y} = 0$, and solve this equation to determine the terminal

velocity: $v = \frac{mgR}{B_0^2 a^2}$.

ANSWER SHEET FOR PHYSICS C PRACTICE TEST II

Section I: Mechanics

1 ⊂A⊃ ⊂B⊃ ⊂C⊃ ⊂D⊃ ⊂E⊃
2 ⊂A⊃ ⊂B⊃ ⊂C⊃ ⊂D⊃ ⊂E⊃
3 ⊂A⊃ ⊂B⊃ ⊂C⊃ ⊂D⊃ ⊂E⊃
4 ⊂A⊃ ⊂B⊃ ⊂C⊃ ⊂D⊃ ⊂E⊃
5 ⊂A⊃ ⊂B⊃ ⊂C⊃ ⊂D⊃ ⊂E⊃
6 ⊂A⊃ ⊂B⊃ ⊂C⊃ ⊂D⊃ ⊂E⊃
7 ⊂A⊃ ⊂B⊃ ⊂C⊃ ⊂D⊃ ⊂E⊃
8 ⊂A⊃ ⊂B⊃ ⊂C⊃ ⊂D⊃ ⊂E⊃
9 ⊂A⊃ ⊂B⊃ ⊂C⊃ ⊂D⊃ ⊂E⊃
10 ⊂A⊃ ⊂B⊃ ⊂C⊃ ⊂D⊃ ⊂E⊃
11 ⊂A⊃ ⊂B⊃ ⊂C⊃ ⊂D⊃ ⊂E⊃
12 ⊂A⊃ ⊂B⊃ ⊂C⊃ ⊂D⊃ ⊂E⊃
13 ⊂A⊃ ⊂B⊃ ⊂C⊃ ⊂D⊃ ⊂E⊃
14 ⊂A⊃ ⊂B⊃ ⊂C⊃ ⊂D⊃ ⊂E⊃
15 ⊂A⊃ ⊂B⊃ ⊂C⊃ ⊂D⊃ ⊂E⊃
16 ⊂A⊃ ⊂B⊃ ⊂C⊃ ⊂D⊃ ⊂E⊃
17 ⊂A⊃ ⊂B⊃ ⊂C⊃ ⊂D⊃ ⊂E⊃
18 ⊂A⊃ ⊂B⊃ ⊂C⊃ ⊂D⊃ ⊂E⊃
19 ⊂A⊃ ⊂B⊃ ⊂C⊃ ⊂D⊃ ⊂E⊃
20 ⊂A⊃ ⊂B⊃ ⊂C⊃ ⊂D⊃ ⊂E⊃
21 ⊂A⊃ ⊂B⊃ ⊂C⊃ ⊂D⊃ ⊂E⊃
22 ⊂A⊃ ⊂B⊃ ⊂C⊃ ⊂D⊃ ⊂E⊃
23 ⊂A⊃ ⊂B⊃ ⊂C⊃ ⊂D⊃ ⊂E⊃
24 ⊂A⊃ ⊂B⊃ ⊂C⊃ ⊂D⊃ ⊂E⊃
25 ⊂A⊃ ⊂B⊃ ⊂C⊃ ⊂D⊃ ⊂E⊃
26 ⊂A⊃ ⊂B⊃ ⊂C⊃ ⊂D⊃ ⊂E⊃
27 ⊂A⊃ ⊂B⊃ ⊂C⊃ ⊂D⊃ ⊂E⊃
28 ⊂A⊃ ⊂B⊃ ⊂C⊃ ⊂D⊃ ⊂E⊃
29 ⊂A⊃ ⊂B⊃ ⊂C⊃ ⊂D⊃ ⊂E⊃
30 ⊂A⊃ ⊂B⊃ ⊂C⊃ ⊂D⊃ ⊂E⊃
31 ⊂A⊃ ⊂B⊃ ⊂C⊃ ⊂D⊃ ⊂E⊃
32 ⊂A⊃ ⊂B⊃ ⊂C⊃ ⊂D⊃ ⊂E⊃
33 ⊂A⊃ ⊂B⊃ ⊂C⊃ ⊂D⊃ ⊂E⊃
34 ⊂A⊃ ⊂B⊃ ⊂C⊃ ⊂D⊃ ⊂E⊃
35 ⊂A⊃ ⊂B⊃ ⊂C⊃ ⊂D⊃ ⊂E⊃

Section I: Electricity and Magnetism

36 ⊂A⊃ ⊂B⊃ ⊂C⊃ ⊂D⊃ ⊂E⊃
37 ⊂A⊃ ⊂B⊃ ⊂C⊃ ⊂D⊃ ⊂E⊃
38 ⊂A⊃ ⊂B⊃ ⊂C⊃ ⊂D⊃ ⊂E⊃
39 ⊂A⊃ ⊂B⊃ ⊂C⊃ ⊂D⊃ ⊂E⊃
40 ⊂A⊃ ⊂B⊃ ⊂C⊃ ⊂D⊃ ⊂E⊃
41 ⊂A⊃ ⊂B⊃ ⊂C⊃ ⊂D⊃ ⊂E⊃
42 ⊂A⊃ ⊂B⊃ ⊂C⊃ ⊂D⊃ ⊂E⊃
43 ⊂A⊃ ⊂B⊃ ⊂C⊃ ⊂D⊃ ⊂E⊃
44 ⊂A⊃ ⊂B⊃ ⊂C⊃ ⊂D⊃ ⊂E⊃
45 ⊂A⊃ ⊂B⊃ ⊂C⊃ ⊂D⊃ ⊂E⊃
46 ⊂A⊃ ⊂B⊃ ⊂C⊃ ⊂D⊃ ⊂E⊃
47 ⊂A⊃ ⊂B⊃ ⊂C⊃ ⊂D⊃ ⊂E⊃
48 ⊂A⊃ ⊂B⊃ ⊂C⊃ ⊂D⊃ ⊂E⊃
49 ⊂A⊃ ⊂B⊃ ⊂C⊃ ⊂D⊃ ⊂E⊃
50 ⊂A⊃ ⊂B⊃ ⊂C⊃ ⊂D⊃ ⊂E⊃
51 ⊂A⊃ ⊂B⊃ ⊂C⊃ ⊂D⊃ ⊂E⊃
52 ⊂A⊃ ⊂B⊃ ⊂C⊃ ⊂D⊃ ⊂E⊃
53 ⊂A⊃ ⊂B⊃ ⊂C⊃ ⊂D⊃ ⊂E⊃
54 ⊂A⊃ ⊂B⊃ ⊂C⊃ ⊂D⊃ ⊂E⊃
55 ⊂A⊃ ⊂B⊃ ⊂C⊃ ⊂D⊃ ⊂E⊃
56 ⊂A⊃ ⊂B⊃ ⊂C⊃ ⊂D⊃ ⊂E⊃
57 ⊂A⊃ ⊂B⊃ ⊂C⊃ ⊂D⊃ ⊂E⊃
58 ⊂A⊃ ⊂B⊃ ⊂C⊃ ⊂D⊃ ⊂E⊃
59 ⊂A⊃ ⊂B⊃ ⊂C⊃ ⊂D⊃ ⊂E⊃
60 ⊂A⊃ ⊂B⊃ ⊂C⊃ ⊂D⊃ ⊂E⊃
61 ⊂A⊃ ⊂B⊃ ⊂C⊃ ⊂D⊃ ⊂E⊃
62 ⊂A⊃ ⊂B⊃ ⊂C⊃ ⊂D⊃ ⊂E⊃
63 ⊂A⊃ ⊂B⊃ ⊂C⊃ ⊂D⊃ ⊂E⊃
64 ⊂A⊃ ⊂B⊃ ⊂C⊃ ⊂D⊃ ⊂E⊃
65 ⊂A⊃ ⊂B⊃ ⊂C⊃ ⊂D⊃ ⊂E⊃
66 ⊂A⊃ ⊂B⊃ ⊂C⊃ ⊂D⊃ ⊂E⊃
67 ⊂A⊃ ⊂B⊃ ⊂C⊃ ⊂D⊃ ⊂E⊃
68 ⊂A⊃ ⊂B⊃ ⊂C⊃ ⊂D⊃ ⊂E⊃
69 ⊂A⊃ ⊂B⊃ ⊂C⊃ ⊂D⊃ ⊂E⊃
70 ⊂A⊃ ⊂B⊃ ⊂C⊃ ⊂D⊃ ⊂E⊃

Physics Formulas

TABLE OF INFORMATION

Constants and Conversion Factors

1 unified atomic mass unit.	$1\ u = 1.66 \times 10^{-27}\ kg = 931\ MeV / c^2$
Proton mass	$m_p = 1.67 \times 10^{-27}\ kg$
Neutron mass	$m_n = 1.67 \times 10^{-27}\ kg$
Electron mass	$m_e = 9.11 \times 10^{-31}\ kg$
Magnitude of electron charge	$e = 1.60 \times 10^{-19}$
Avogadro's number	$N_0 = 6.02 \times 10^{23}\ mol^{-1}$
Universal gas constant	$R = 8.31\ J / (mol \times K)$
Boltzmann's constant	$k_B = 1.38 \times 10^{-23}\ J/K$
Speed of light	$c = 3.00 \times 10^8\ m/s$
Planck's constant	$h = 6.63 \times 10^{-34} J \times s = 4.14 \times 10^{-15}\ eV \times s$
	$Hc = 1.99 \times 10^{-25} J \times m = 1.24 \times 10^3\ eV \times nm$
Vacuum permittivity	$e_0 = 8.85 \times 10^{-12} C2/N \times m^2$
Coulomb's law constant	$k = 1/4\pi e_0 = 9.0 \times 10^9 N \times m^2/C^2$
Vacuum permeability	$m_0 = 4p \times 10^{-7} Wb/(A \times m)$
Magnetic constant	$k' = m_0/4\pi \times 10^{-7} Wb/(A \times m)$
Universal gravitational constant	$G = 6.67 \times 10^{-11}\ m^3/kg \times s^2$
Acceleration due to gravity at the Earth's surface	$g = 9.8\ m/s^2$
1 atmosphere pressure	$1\ atm = 1.0 \times 10^5\ N/m^2 = 1.0 \times 10^5\ Pa$
1 electron volt	$1\ eV = 1.60 \times 10^{-19} J$
1 angstrom	$1\ Å = 1 \times 10^{-10}\ m$

Units

Name	Symbol
meter	m
kilogram	kg
second	s
ampere	A
kelvin	K
mole	mol
hertz	Hz
newton	N
pascal	Pa
joule	J
watt	W
coulomb	C
volt	V
ohm	W
henry	H
farad	F
weber	Wb
tesla	T
degree Celsius	°C
electron-volt	eV

Prefixes

Factor	Prefix	Symbol
10^9	giga	G
10^6	mega	M
10^3	kilo	k
10^{-2}	centi	c
10^{-3}	milli	m
10^{-6}	micro	m
10^{-9}	nano	n
10^{-12}	pico	p

Values of Trigonometric Functions For Common Angles

Angle	Sin	Cos	Tan
0°	0	1	0
30°	1 / 2	$\sqrt{3}/2$	$\sqrt{3}/3$
37°	3 / 5	4 / 5	3 / 4
45°	$\sqrt{2}/2$	$\sqrt{2}/2$	1
53°	4 / 5	3 / 5	4 / 3
60°	$\sqrt{3}/2$	1 / 2	$\sqrt{3}$
90°	1	0	∞

Newtonian Mechanics

a = acceleration	F = force
f = frequency	h = height
J = impulse	K = kinetic energy
k = spring constant	l = length
m = mass	N = normal force
P = power	p = momentum
r = radius or distance	s = displacement
T = period	t = time
U = potential energy	v = velocity or speed
W = work	x = position

Physics C

PRACTICE TEST 2

SECTION I—MECHANICS

Directions: Each question listed below has five possible choices. Select the best answer given the information in each problem, and mark the corresponding oval on the answer sheet. (You may assume $g = 10$ m/s^2).

Questions 1 and 2 refer to the following figure.

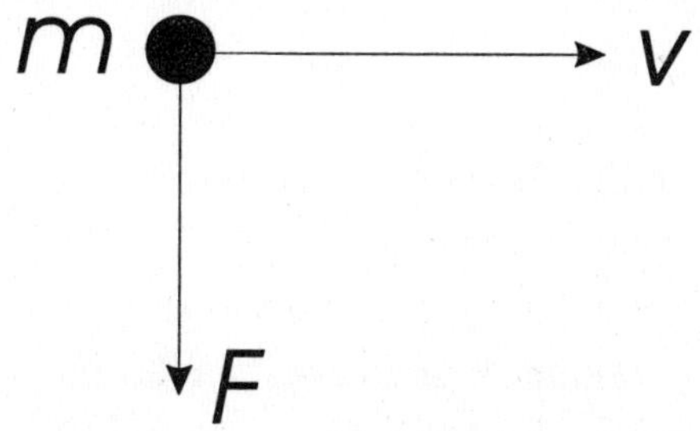

1. At the instant depicted in the figure, the instantaneous power of the force acting on the mass is

(A) 0.
(B) $\frac{Fv}{m}$.
(C) Fv.
(D) mvF.
(E) $2Fv$.

2. At times later than depicted in the figure, the kinetic energy of the mass

(A) decreases.
(B) remains constant.
(C) increases.
(D) decreases, then increases.
(E) increases, then decreases.

3. A cannon fires a projectile horizontally at 200m/s from the edge of an 80m-high cliff. How long is the projectile in flight before hitting the ground. Neglect air resistance.

(A) 2 sec
(B) 4 sec
(C) 8 sec
(D) 12 sec
(E) 16 sec

4. A particle is moving in one dimension with its position as a function of time given by $x(t)$. The initial velocity, $v(0)$, is positive, while the acceleration, a, is constant and negative. For $\frac{t >> v(O)}{a}$, $v(t)$

(A) approaches 0.
(B) is $v(0)$.
(C) is $-v(0)$.
(D) is positive and increasing.
(E) is negative and decreasing.

5. Jack throws a ball to Dave, who catches it below the height from which Jack threw it. Neglecting friction, how do the vertical components of the ball's velocity compare at Jack's (v_J) and Dave's (v_D) locations? Upward components are defined as positive.

(A) $v_J = -v_D$
(B) $v_J = v_D$
(C) $v_J > v_D$
(D) $v_J < v_D$
(E) $v_J < -v_D$

6. A block slides down a frictionless inclined plane. The horizontal component of the velocity of the block

(A) does not change.
(B) increases because gravity exerts a force on the block.
(C) decreases because gravity exerts a force on the block.
(D) increases because the inclined plane exerts a force on the block.
(E) decreases because the inclined plane exerts a force on the block.

7. A particle is moving in one dimension whose position as a function of time is described by $x(t) = t^3 + 3t^2 + 7$. What is the acceleration when the velocity is zero?

(A) 0
(B) 3
(C) 6
(D) –6
(E) either 6 or –6

8. A mass is at rest on a horizontal surface with friction. The net force on the mass is

(A) zero because gravity is balanced by friction.
(B) zero because friction is balanced by the normal force of the surface on the mass.
(C) zero because gravity is balanced by the normal force of the surface on the mass.
(D) down because of gravity.
(E) horizontal because of friction.

9. A mass slides on a frictionless, horizontal surface under the action of a horizontal force. After the force stops acting on the mass, its velocity

(A) gradually decreases to zero.
(B) remains constant.
(C) immediately decreases to zero.
(D) continues to increase, then levels off.
(E) cannot be determined.

10. A particle moves in one dimension according to the equation $x(t) = t^2 + 5t + 2$. The acceleration and velocity are both zero when t is

(A) 0.
(B) 5.
(C) $\frac{5}{2}$.
(D) $\frac{2}{5}$.
(E) The acceleration is never zero.

11. Two masses collide while sliding on a frictionless table and stick together. Which of the following is the same before and after collision?

(A) Total momentum
(B) Total kinetic energy
(C) Total velocity
(D) Both (A) and (B)
(E) Both (B) and (C)

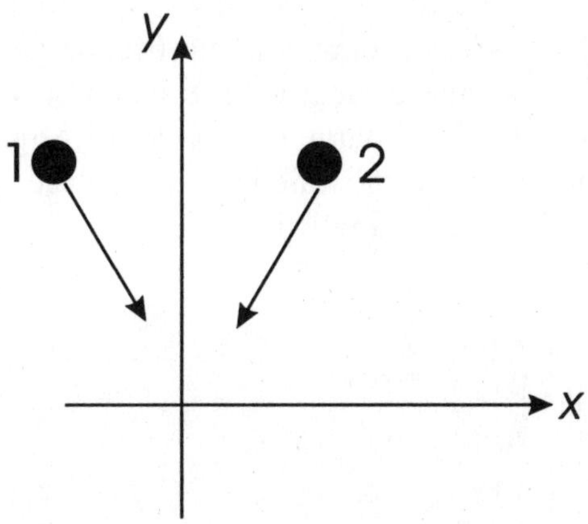

12. Two masses with equal speed, v, are moving towards the origin of coordinates, as depicted in the figure above. Their velocity vectors are symmetrically oriented about the y-axis with nonzero y components. A third equal mass will collide with the first two at the origin, where all three will stick together and remain at rest. The velocity of the third mass must have magnitude

(A) $2v$.
(B) between v and $2v$.
(C) between 0 and v.
(D) between 0 and $2v$.
(E) None of these

13. A cube of uniform density is resting on the table. The cube is pivoted on an edge such that the center of mass is directly above the pivot. The cube

(A) is in stable equilibrium.
(B) is in unstable equilibrium.
(C) has a gravitational torque acting on it.
(D) can be disturbed slightly without tipping over.
(E) has no gravitational forces acting on it.

14. Particle A is raised to a height $2h$, then lowered to height h. Particle B is raised directly to height h. Both particles have the same mass and started at height 0. If friction is neglected, the work required

(A) is greater for A than for B.
(B) is greater for B than for A.
(C) is equal for A and B.
(D) depends on the velocity of the motion.
(E) depends on the acceleration of the motion.

15. A ball of mass m is dropped from height h on a spring with elastic constant k. Neglecting the mass of the spring, what is the maximum compression of the spring?

(A) $mghk$
(B) $\sqrt{mghk}$
(C) $\frac{mghk}{2}$
(D) $\sqrt{\frac{mghk}{2}}$
(E) $\sqrt{\frac{2mgh}{k}}$

16. A bead is released from the rim of a hemispherical bowl of radius R. If friction is neglected, what is the speed at the bottom of the bowl?

(A) $2gR$

(B) $\sqrt{2gR}$

(C) $\sqrt{2mgR}$

(D) $\sqrt{\frac{gR}{2}}$

(E) $\sqrt{\frac{mgR}{2}}$

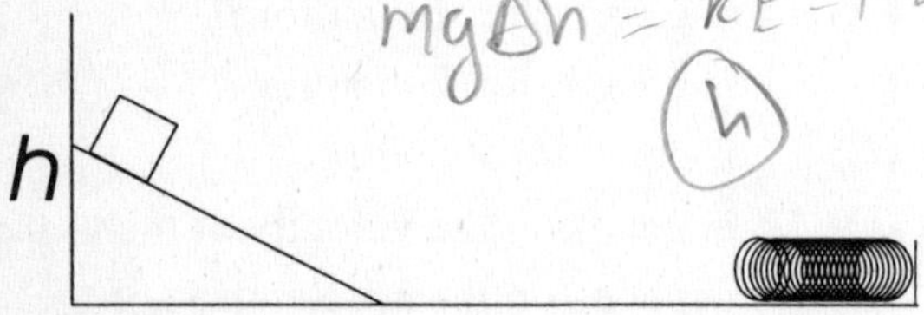

17. A block slides down an inclined plane of height h to collide with a spring, as shown in the figure above. After rebounding from the spring, to what height does the block rise? Neglect friction and the mass of the spring.

(A) $\frac{\sqrt{3}h}{2}$

(B) $\frac{3h}{4}$

(C) h

(D) The height depends on the angle of the incline.

(E) The height depends on the spring's elastic constant.

18. Donny and Marie are sitting on a boat of negligible mass, separated by 1 meter. Donny has a mass of 75kg. When they switch places, the boat moves $\frac{1}{7}$ m in the direction that Donny moves. What is Marie's mass?

(A) 50 kg

(B) 60 kg

(C) 75 kg

(D) 100 kg

(E) 125 kg

19. A 5kg block is sliding, without friction, along a horizontal surface at 2m/s. A force of 10N is applied for 2 sec perpendicular to the original motion. What is the kinetic energy of the block after the force is applied?

(A) 5 J

(B) 10 J

(C) 20 J

(D) 40 J

(E) 50 J

20. A 100g ball bounces off a wall elastically. Its initial speed is 3m/s perpendicular to the wall. If the collision lasts for 10ms, what is the average force exerted by the wall on the ball?

(A) 3N

(B) 10N

(C) 30N

(D) 60N

(E) 100N

21. A 5kg mass moving at 2m/s collides with a 15kg mass at rest. After the collision, they stick together. What is the speed of the combined mass?

(A) 0.5m/s
(B) 1m/s
(C) 2m/s
(D) 5m/s
(E) 10m/s

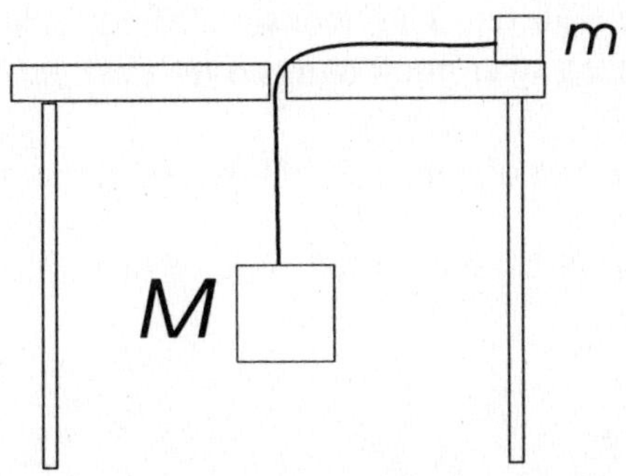

22. A mass m is free to move on a horizontal, frictionless table. It is attached to a hanging mass M by a massless string through a hole in the table, as shown in the figure above. If m = 2kg and M = 8kg, what tangential velocity must be imparted to m so that it will revolve around the hole in a circle of 40cm radius?

(A) 1m/s
(B) 2m/s
(C) 4m/s
(D) 8m/s
(E) 10m/s

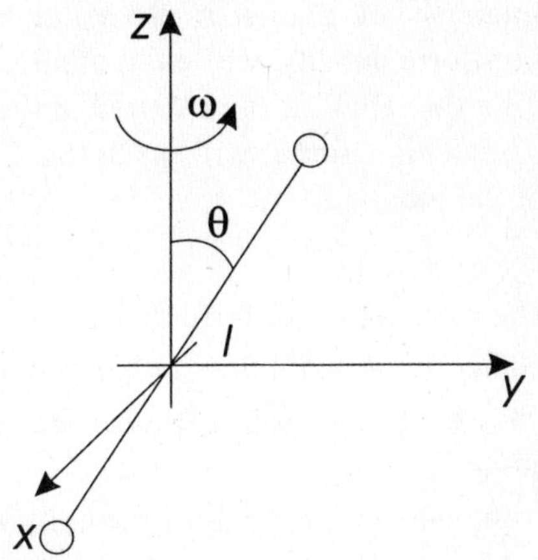

23. Consider a massless rod of length l with a mass m at each end. The assembly rotates about the z-axis with angular speed ω, as shown in the figure above. The rod is inclined to the z-axis at an angle θ. The angular momentum of the system is

(A) $\frac{ml^2\omega}{2}$

(B) $\frac{ml^2\omega}{2\sin^2\theta}$

(C) $\frac{ml^2\omega}{2\sin\theta}$

(D) $ml^2\omega$

(E) $ml^2\omega\sin\theta$

24. Two cylinders roll without friction down an inclined plane. Their diameters are equal, but one is of uniform density, while the other consists of a thin shell. If they started at the same time, which cylinder will reach the bottom of the incline first?

(A) The cylinder of uniform density
(B) The cylinder with mass in a thin shell
(C) They both reach the bottom at the same time
(D) It depends on the mass of the cylinders
(E) It depends on the diameter of the cylinders

25. A roller coaster has a circular loop of radius R. What should the speed, v, of the car be at the top of the loop, so the riders feel weightless?

(A) $2gR$
(B) gR
(C) $\sqrt{2gR}$
(D) $\sqrt{gR}$
(E) $\sqrt{\frac{gR}{2}}$

26. A 200g cylindrical shell of 50cm radius is rotating about its axis at 30 rad/sec. What is the final angular speed after a torque of 2 N–m is applied for 2 sec?

(A) 60 rad/s
(B) 80 rad/s
(C) 110 rad/s
(D) 160 rad/s
(E) 300 rad/s

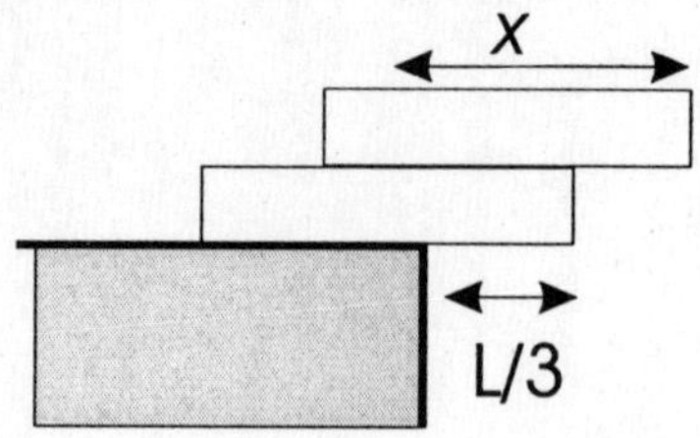

27. Two books of equal mass and uniform density are placed at the edge of a table, as shown in the figure above. The books are of equal length, L, and the lower book hangs over the edge of the table at a distance $\frac{L}{3}$. How far, x, can the upper book hang over the edge of the table?

(A) $\frac{L}{6}$
(B) $\frac{L}{3}$
(C) $\frac{L}{2}$
(D) $\frac{2L}{3}$
(E) $\frac{5L}{6}$

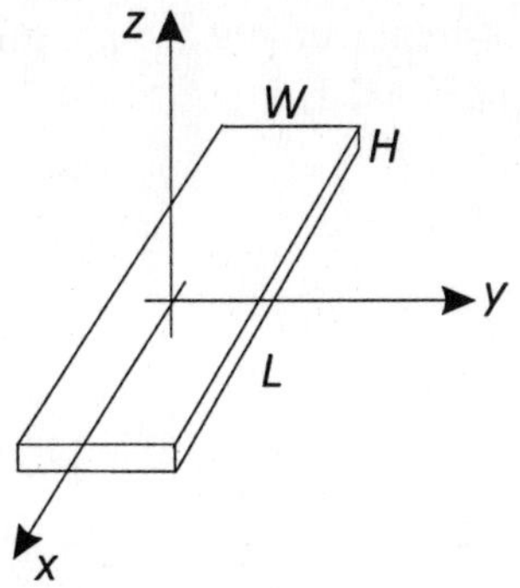

28. Consider a rectangular solid of dimensions $L \times W \times H$ and uniform density, as shown in the figure above. The dimensions are such that $L > W > H$. For a fixed applied torque along an axis, which axis will result in the highest angular acceleration?

(A) x
(B) y
(C) z
(D) All equal
(E) Cannot be determined

29. A simple harmonic oscillator's position is given by $x(t) = A\cos(\omega t)$. The mechanical energy of the oscillator is greatest when t is

(A) 0
(B) $\frac{1}{\omega}$
(C) $\frac{\pi}{\omega}$
(D) $\frac{\pi}{(2\omega)}$
(E) None of the above

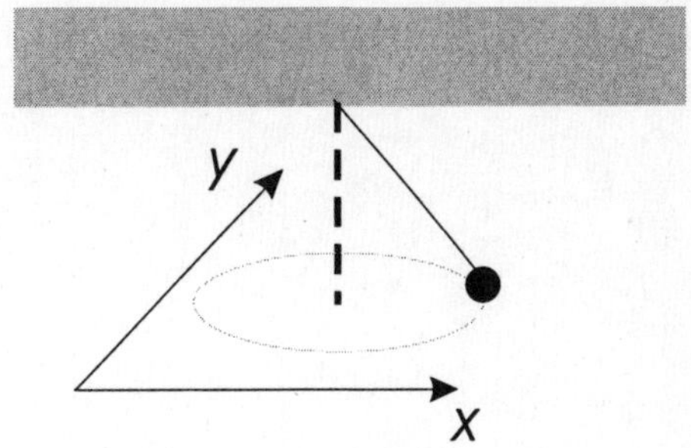

30. A pendulum is free to swing along x and y, as shown in the figure above. The period of small oscillations

(A) depends on the relative amplitudes of the x and y motion.
(B) is greater than the period of oscillation in a plane.
(C) is greatest for circular motion in the x-y plane.
(D) is independent of the relative amplitude.
(E) is smallest for circular motion in the x-y plane.

31. A child is swinging from a rope of length L. If she starts her swing when the rope is horizontal, what is the acceleration at the bottom of the swing?

(A) $\frac{g}{2}$
(B) g
(C) $2g$
(D) $\frac{2}{g}$
(E) $\frac{2g}{L}$

32. A satellite is in a circular orbit around the Earth at a height that is equal to the radius of the Earth. The acceleration of the satellite is

(A) 0
(B) $2g$
(C) g
(D) $\frac{g}{2}$
(E) $\frac{g}{4}$

33. A block of mass *M*, attached to a spring, is oscillating with an amplitude *A*. As the block passes through its equilibrium position, an equal mass *M* is dropped on the block, to which it sticks. The subsequent amplitude of the oscillation is

(A) $\frac{A}{2}$
(B) $\frac{A}{\sqrt{2}}$
(C) A
(D) $\sqrt{2A}$
(E) $2A$

34. A single planet is in orbit around a massive star. Which of the following quantities is conserved by the planet? Assume that the planet's mass is negligible compared to the star's mass.

(A) Kinetic energy
(B) Potential energy
(C) Linear momentum
(D) Angular momentum
(E) Gravitational energy

35. A mass oscillates on a spring with amplitude *A*. At what displacement from equilibrium is the P.E. equal to K.E.?

(A) 0
(B) $\frac{A}{\sqrt{2}}$
(C) $\frac{A}{2}$
(D) $A\sqrt{\frac{3}{2}}$
(E) A

SECTION I—ELECTRICITY AND MAGNETISM

For the following problems on electrostatics, the Coulomb Law constant, $\frac{1}{(4\pi\varepsilon_0)}$, is denoted by *k*.

36. Three charges of magnitude *q* are positioned at the vertices of an equilateral triangle of side *d*. How much energy was required to assemble these charges if they were originally at infinity?

(A) $\frac{kq^2}{d^2}$
(B) $\frac{kq^2}{d}$
(C) $\frac{2kq^2}{d^2}$
(D) $\frac{2kq^2}{d}$
(E) $\frac{3kq^2}{d}$

37. Two concentric spheres of radii a and b, respectively, are uniformly charged with Q and $-Q$, respectively. The electric field at a distance $r>a>b$ is

(A) 0.

(B) $kQ\left(\frac{1}{a^2}-\frac{1}{b^2}\right)$.

(C) $kQ\left(\frac{1}{b^2}-\frac{1}{a^2}\right)$.

(D) $\frac{kQ}{r}\left(\frac{1}{b}-\frac{1}{a}\right)$.

(E) $\frac{kQ}{r^2}$.

38. The plates of a parallel-plate capacitor are charged to $+Q$ and $-Q$, while the space between the plates contains a vacuum. A material of dielectric constant $K_e>1$ is introduced into the space between the plates. The energy stored in the capacitor changes by a factor of

(A) K_e.

(B) K_e^2.

(C) $\frac{1}{K_e}$.

(D) $\frac{1}{K_e^2}$.

(E) Does not change

39. Two long, concentric cylindrical conductors have equal and opposite charges. The inner cylinder is charged $+Q$ and has radius a, while the outer cylinder is charged $-Q$ has radius b. They are both of length $L>>b$. The electric field at a distance from the axis $a<r<b$ is given by

(A) 0.

(B) $\frac{2kQ}{(rL)}$

(C) $\frac{kQ}{L}\left(\frac{1}{a}-\frac{1}{b}\right)$.

(D) $kQ\left(\frac{1}{a^2}\right)$.

(E) $kQ\left(\frac{1}{a^2}-\frac{1}{b^2}\right)$.

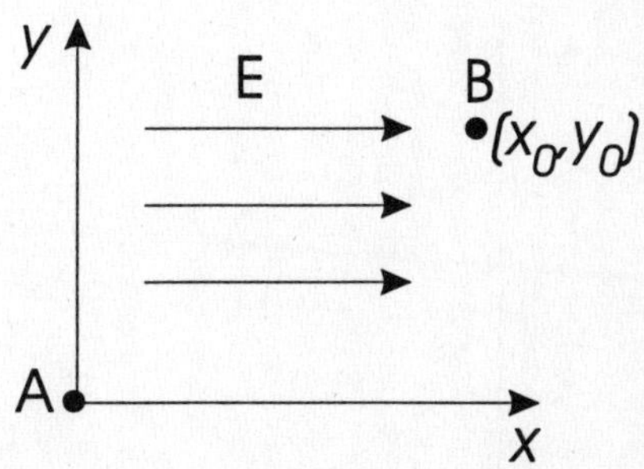

40. A charge Q is moved slowly from A to B in a region of uniform field E, as shown in the figure above. Point A is at the origin of coordinates, while B has coordinates (x_0, y_0). The change in energy of the particle is

(A) $QE\sqrt{x_0^2+y_0^2}$.

(B) $QE(x_0^2+y_0^2)$.

(C) $Q^2E\sqrt{x_0^2+y_0^2}$.

(D) QEx_0.

(E) QEy_0.

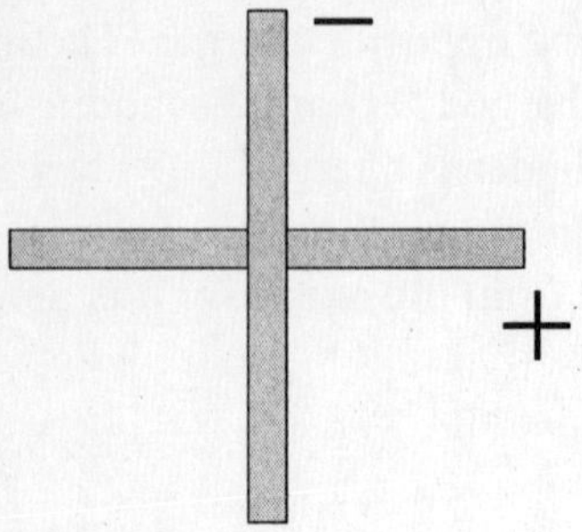

41. Two perpendicular infinite plates are uniformly, equally, and oppositely charged, as shown in the figure above. The electric field is best represented by

(A) 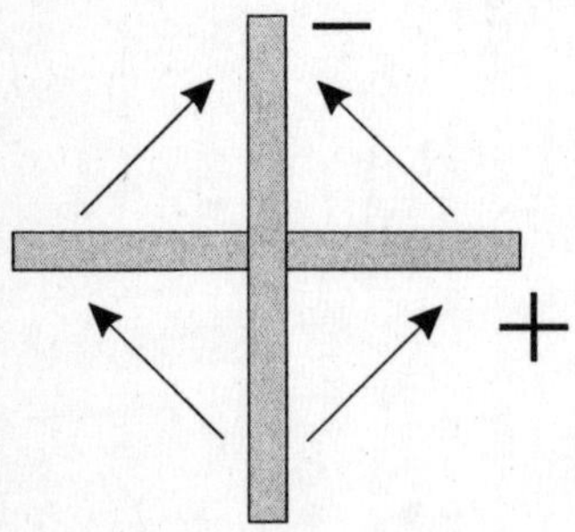

(B) 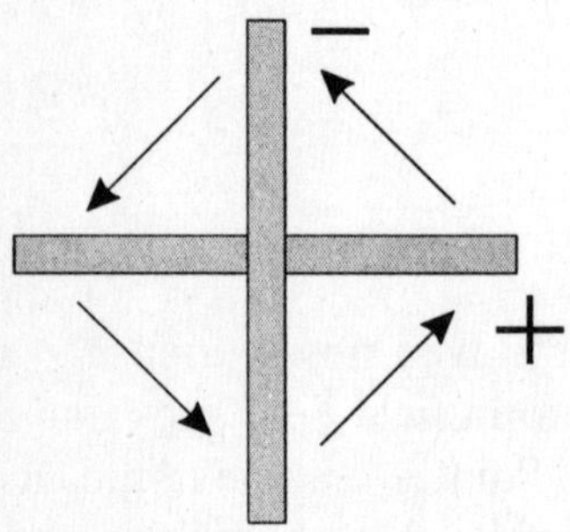

(C) 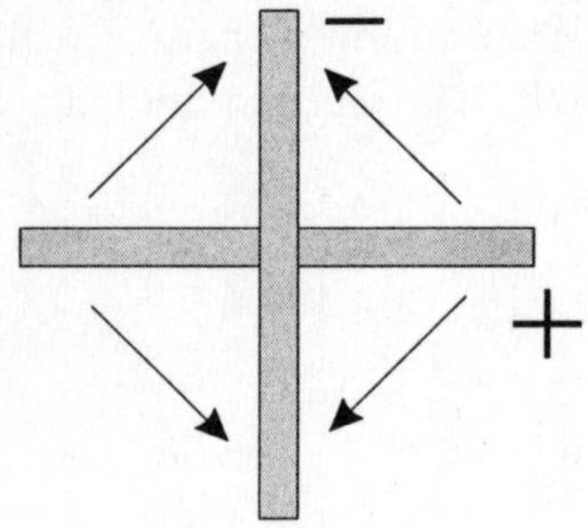

(D) 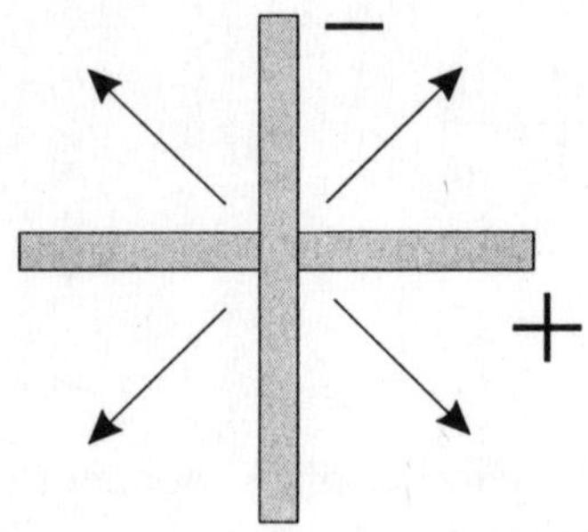

(E) zero everywhere.

42. A sphere of radius R is uniformly charged on its surface with total charge Q. The work required to bring a charge q from infinity to the center of the sphere is

(A) 0.

(B) $\frac{kqQ}{R}$.

(C) $\frac{kQ}{r}$.

(D) $\frac{kq}{R}$.

(E) infinity.

43. Two equal and opposite charges, Q and $-Q$, are placed in the x-y plane at $(b, 0)$ and $(-b, 0)$, respectively. The work required to bring a charge q from infinity to the origin is

(A) 0.

(B) $\frac{kQq}{b}$.

(C) $\frac{2kQq}{b}$.

(D) $\frac{kQq}{(2b)}$.

(E) $\frac{kQq}{b^2}$.

44. Two equal charges, Q, are located at a distance, b, apart. Another charge q is placed at the midpoint between them. For what value of q (in terms of Q) is the system in equilibrium?

(A) $q = Q$

(B) $q = -Q$

(C) $q = -\frac{1}{2}Q$

(D) $q = 4Q$

(E) $q = -\frac{1}{4}Q$

45. Charge Q is uniformly distributed on the surface of a hollow sphere of radius R. How much energy is required to move a charge q from the center to the surface of the sphere?

(A) 0

(B) $\frac{kQq}{R}$

(C) $\frac{kQq}{R^2}$

(D) $\frac{2kQq}{R}$

(E) $\frac{2kQq}{R^2}$

46. Two conducting parallel plates, separated by d, are charged to $+Q$ and $-Q$, respectively. When a dielectric material is inserted between the plates, the potential difference between the plates

(A) increases.

(B) decreases.

(C) remains constant.

(D) increases or decreases depending on the material.

(E) increases as the material is inserted, then decreases.

47. Two masses, m and M, have charges q and Q, respectively. Both particles are accelerated from rest by two charged parallel plates with potential difference V. If $M = 2m$ and $Q = 2q$, the ratio of the work done on of mass M to that on m is

(A) $\frac{1}{4}$.

(B) $\frac{1}{2}$.

(C) 1.

(D) 2.

(E) 4.

48. Two masses, m, each with charge q, are traveling toward each other with relative speed v. Their distance of closest approach is

(A) $\frac{2kq}{(mv^2)}$

(B) $\frac{2kq^2}{(mv^2)}$

(C) $\frac{4kq}{(mv^2)}$

(D) $\frac{4kq^2}{(mv^2)}$

(E) $\frac{4kq^2}{(mv)}$

49. A conducting sphere of radius R has charge Q. The electric potential at infinity is defined to be zero. Which of the following is zero inside the sphere?

(A) Potential

(B) Electric field

(C) Gradient of the potential

(D) Both (A) and (B)

(E) Both (B) and (C)

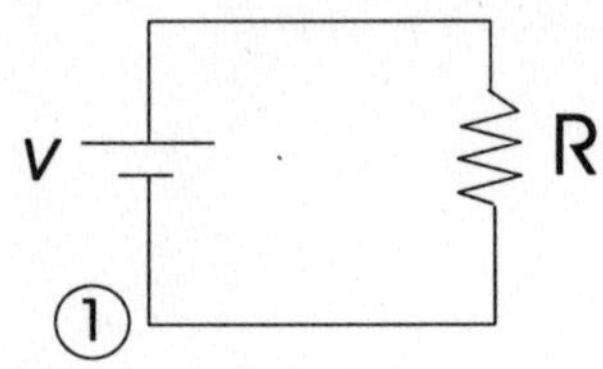

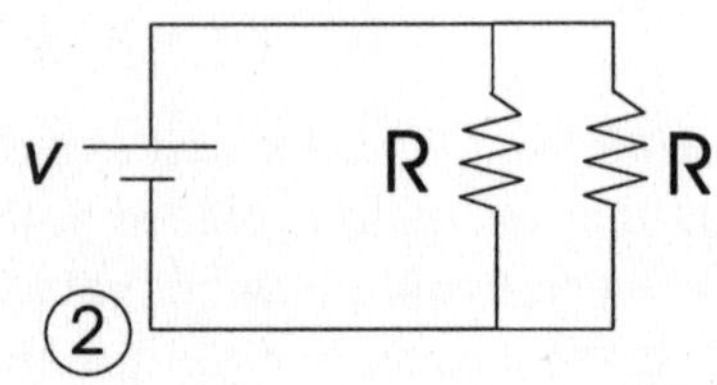

50. Consider the two circuits depicted above. If the internal resistance of the battery is neglected, which of the following is true?

(A) Both dissipate the same power.

(B) Circuit 1 dissipates more power.

(C) Circuit 2 dissipates more power.

(D) Circuit 1 draws more current from the battery.

(E) The voltage across each resistor is lower in circuit 2.

51. A resistor R is connected to the terminals of a battery, resulting in a current flow of 1A. Another equal resistor is added, parallel with the first. The total current flowing form the battery increases to 1.5A. The internal resistance of the battery is

(A) 0.

(B) $\frac{R}{2}$.

(C) R.

(D) $2R$.

(E) $3R$.

52. Two equal resistors are connected to a battery, either in parallel or in series. What is the ratio of the total power dissipation in the parallel divided by the series configuration?

(A) $\frac{1}{2}$

(B) 1

(C) 2

(D) 4

(E) 8

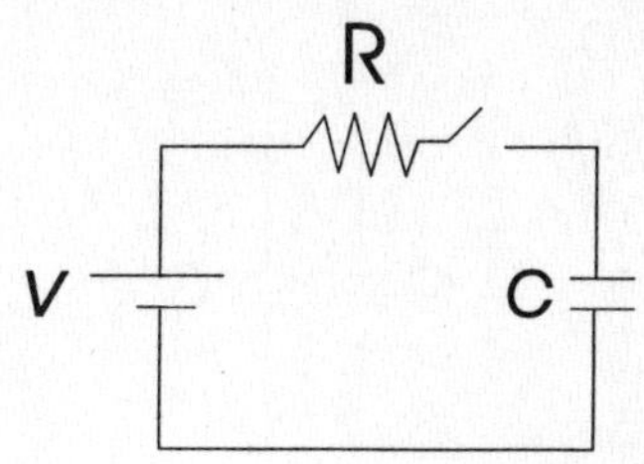

53. When the switch is closed in the circuit depicted above, the current

(A) is maximum initially, decreasing monotonically to zero.

(B) increases gradually to a maximum and then decreases.

(C) increases to an asymptotic value.

(D) remains zero.

(E) decreases, then increases to a constant value.

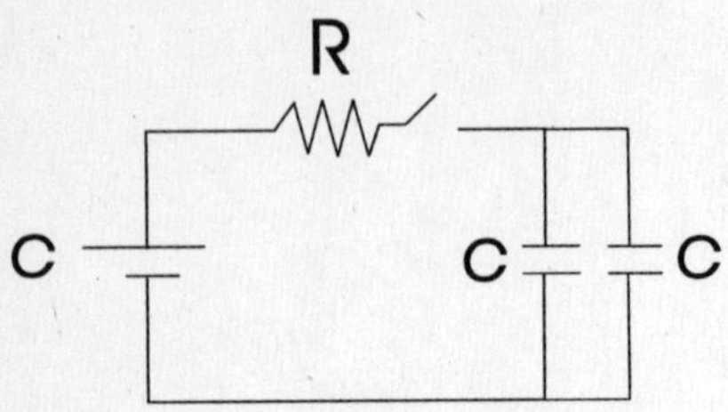

54. Three identical capacitors and a resistor are connected, as depicted in the figure above. Before the switch is closed, there is charge Q on the capacitor on the left and zero charge on the others. A long time after the switch is closed, what is the voltage across the capacitor on the left?

(A) $\frac{Q}{C}$

(B) $\frac{Q}{(3C)}$

(C) $\frac{Q}{(2C)}$

(D) CQ

(E) $\frac{CQ}{3}$

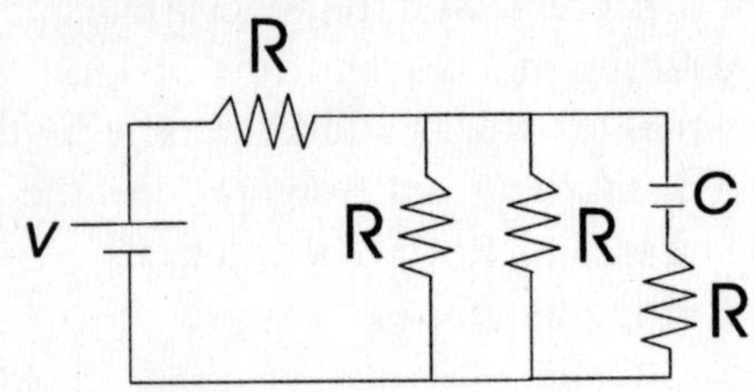

55. The circuit above consists of a battery, four identical resistors, R, and a capacitor, C. What is the current delivered by the battery?

(A) $\frac{V}{R}$

(B) $\left(\frac{3}{4}\right)\frac{V}{R}$

(C) $\left(\frac{2}{3}\right)\frac{V}{R}$

(D) $\left(\frac{1}{2}\right)\frac{V}{R}$

(E) $\left(\frac{1}{3}\right)\frac{V}{R}$

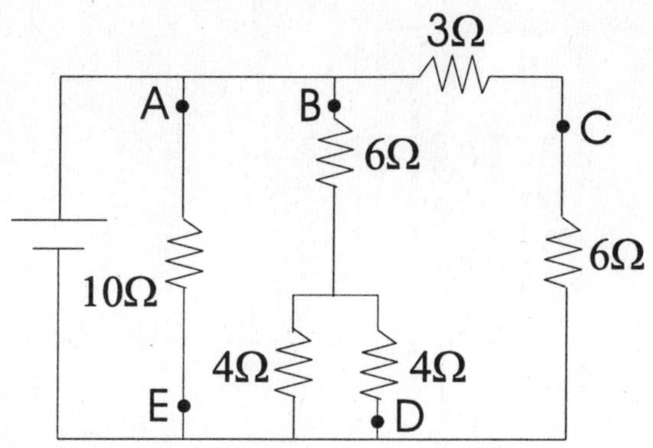

56. In the circuit depicted above, which point has the highest current?

(A) A
(B) B
(C) C
(D) D
(E) E

57. Consider a circuit that is powered by a battery of negligible internal resistance. If the battery is replaced by a battery with nonzero internal resistance, which of the following statements is true?

(A) The power delivered to the circuit is lower.
(B) The power delivered to the circuit is higher.
(C) The current delivered to the battery is higher.
(D) The voltage applied to the circuit is the same.
(E) The voltage applied to the circuit is higher.

58. A positive charge is moving parallel to a long current-carrying wire. If the charge is moving in the same direction as the flow of positive current, the force of the wire on the charge is

(A) zero.
(B) parallel to the wire, opposing the motion of the charge.
(C) parallel to the wire, in the same direction as the motion of the charge.
(D) towards the wire.
(E) away from the wire.

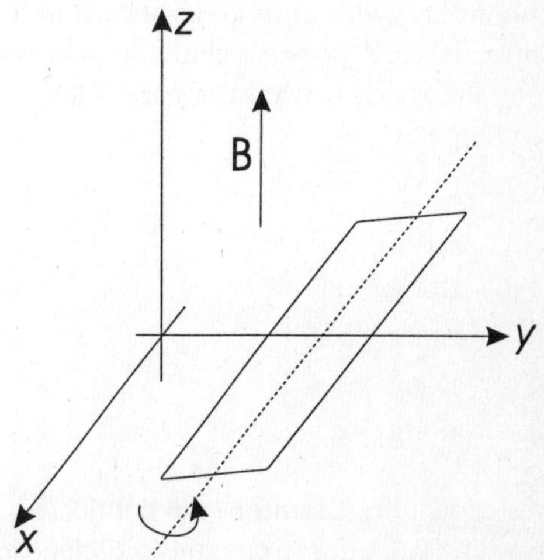

59. A rectangular, conducting loop is rotating at a constant rate about an axis parallel to *x*, as shown in the figure above. There is a uniform magnetic field parallel to the *z*-axis. The emf induced in the loop is

(A) maximum when the loop is parallel to the *x-y* plane.
(B) maximum when the loop is parallel to the *x-z* plane.
(C) minimum when the loop is parallel to the *y-z* plane.
(D) constant.
(E) zero.

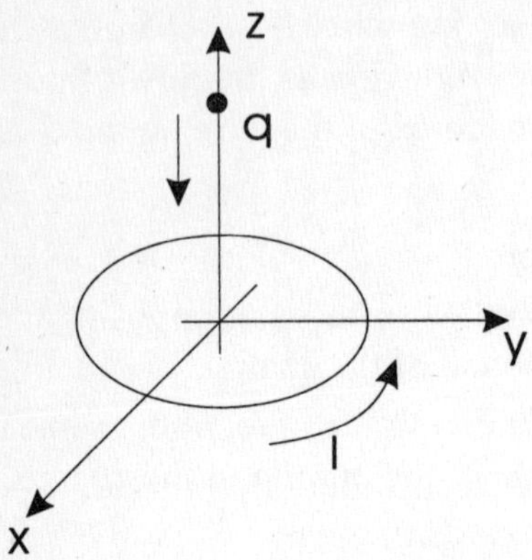

60. A circular wire in the *x-y* plane is carrying a counterclockwise current, as shown in the figure above. A positive charge *q* is moving along the *z*-axis toward the wire. The force on the charge is

(A) zero.
(B) along $+z$.
(C) along $-z$.
(D) along $+x$.
(E) along $-x$.

61. Two capacitors (*A* and *B*) are connected in parallel. Capacitor *A* contains a dielectric material, while capacitor *B* contains a vacuum. The capacitors are connected to a battery until they are charged, then disconnected. When the dielectric material is moved from *A* to *B*,

(A) no current flows.
(B) current flows from *A* to *B*.
(C) current flows from *B* to *A*.
(D) current flows from *A* to *B*, then back to *A*.
(E) current flows from *B* to *A*, then back to *B*.

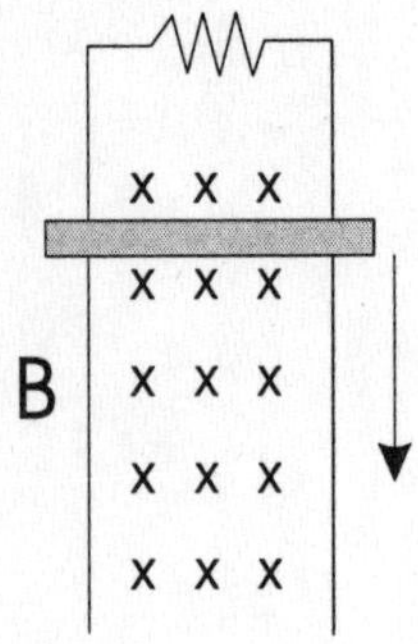

62. A metal bar of width *w* is released from rest near the top of vertical, conducting rails. It falls under the action of gravity, as shown in the figure above. There is a uniform magnetic field perpendicular to the plane of the rails. The circuit is completed at the top by a resistor. If friction is neglected, the bar

(A) accelerates at a constant rate.
(B) falls at a constant speed.
(C) slows to a stop.
(D) accelerates at an increasing rate.
(E) approaches a constant velocity.

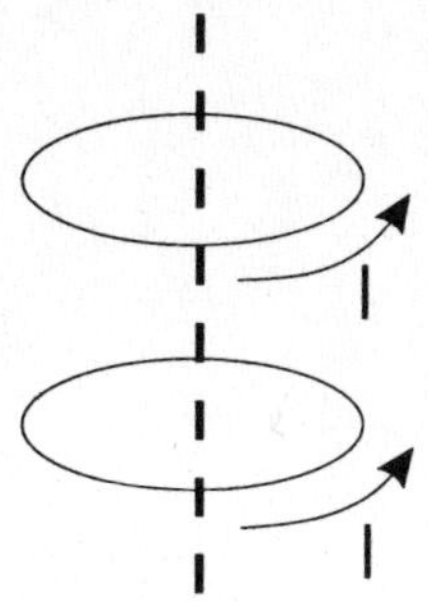

63. Two identical, circular current loops are positioned one directly above the other, as shown in the figure above. The direction of current flow is the same for both loops. The force of one loop on the other

(A) is repulsive.
(B) is attractive.
(C) is zero.
(D) results in a torque on the other loop.
(E) compresses the other loop.

64. There is an infinite current sheet in the x-y plane, with current flowing in the $+y$ direction. A negative charge is moving in the $-y$ direction. The force on the charge is

(A) zero.
(B) in the $+z$ direction.
(C) in the $-z$ direction.
(D) in the $+x$ direction.
(E) in the $-x$ direction.

65. A capacitor and inductor are connected in parallel. A current is made to flow in the circuit, resulting in oscillation. When the current is maximum, the circuit's energy is

(A) zero.
(B) maximum.
(C) stored in the electric field.
(D) stored in the magnetic field.
(E) both (B) and (C).

66. A region of space is enclosed by a conductor. There are no charges or currents in the region. This means that

(A) the electric field must be zero.
(B) the magnetic field must be zero.
(C) both (A) and (B) are true.
(D) the magnetic field is nonzero if the electric field is nonzero.
(E) the electric field is constant.

67. A charged particle is moving in a region of uniform magnetic field in the $+z$ direction. The work done by the field on the particle is

(A) zero if the particle moves in the x-y plane.
(B) zero if the particle moves in the y-z plane.
(C) zero if the particle moves in the x-z plane.
(D) both (B) and (C).
(E) Always zero.

68. Two long, current-carrying wires are oriented perpendicular to each other. The magnetic force of one on the other

(A) is zero.
(B) is attractive.
(C) is repulsive.
(D) produces a torque on the wires.
(E) is independent of the distance between the wires.

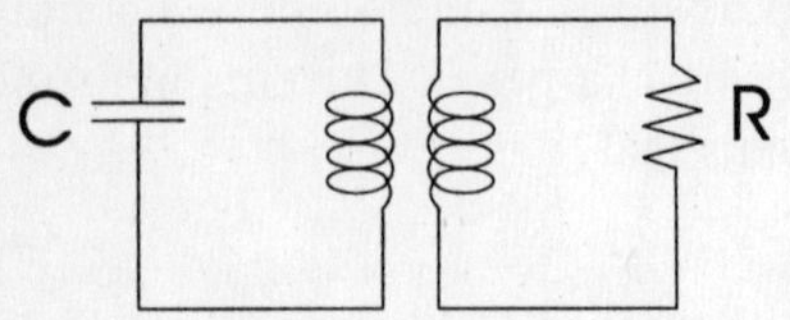

69. An inductor and a capacitor are connected in parallel, as shown in the left half of the figure above. When a second inductor (right) is brought near the inductor in the circuit on the left, the resonant frequency of the circuit

(A) decreases.
(B) increases.
(C) remains the same.
(D) increases or decreases, depending on the value of *R*.
(E) increases or decreases, depending on the value of *C*.

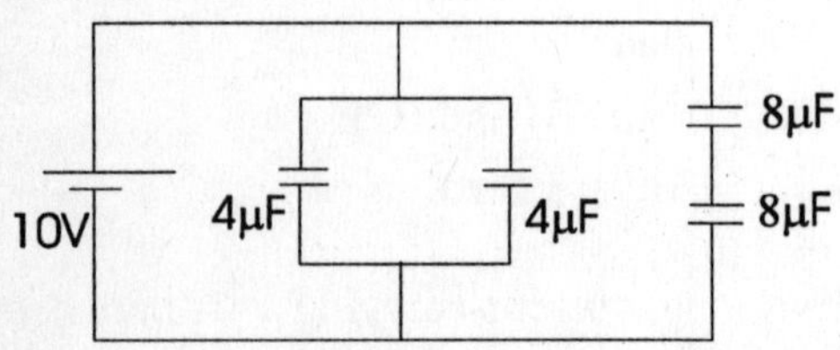

70. Four capacitors are connected to a 10V battery, as shown in the figure above. What is the total charge stored on all the capacitors?

(A) 20 μC
(B) 40 μC
(C) 80 μC
(D) 160 μC
(E) 320 μC

ANSWER SHEET FOR PHYSICS C PRACTICE TEST II

Section II: Mechanics

1 ⊂A⊃ ⊂B⊃ ⊂C⊃ ⊂D⊃ ⊂E⊃ 2 ⊂A⊃ ⊂B⊃ ⊂C⊃ ⊂D⊃ ⊂E⊃ 3 ⊂A⊃ ⊂B⊃ ⊂C⊃ ⊂D⊃ ⊂E⊃

Section II: Electricity and Magnetism

1 ⊂A⊃ ⊂B⊃ ⊂C⊃ ⊂D⊃ ⊂E⊃ 2 ⊂A⊃ ⊂B⊃ ⊂C⊃ ⊂D⊃ ⊂E⊃ 3 ⊂A⊃ ⊂B⊃ ⊂C⊃ ⊂D⊃ ⊂E⊃

SECTION II—MECHANICS

Directions: Answer all three questions. You will have 45 minutes in which to answer all of the questions. Note that each part within a question may not have equal weight.

Mech-1

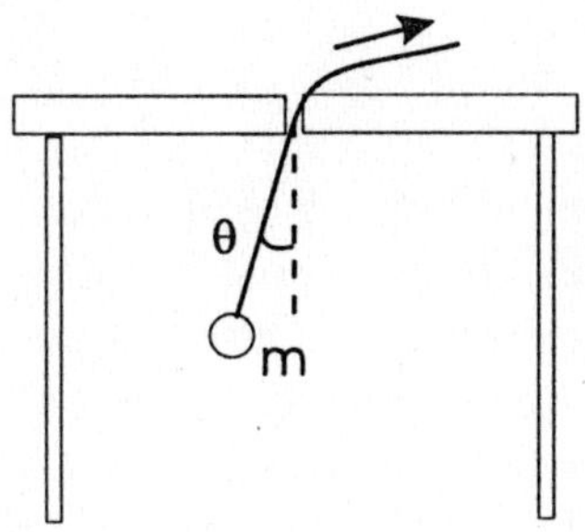

A mass m is swinging from a massless string of initial length y_0 that passes through a hole in the table, as shown in the figure above. The string is pulled up so its length is $y = \frac{y_0}{2}$ when the mass is at the bottom of its swing. The initial angular amplitude is $\theta << 1$. Neglect the time required for the string to be pulled up.

(a) Find the change in gravitational potential energy of the mass when the string is pulled.

(b) Find the speed of the mass immediately after it is pulled up as a function of y_0 and θ_0.

(c) Find the change in kinetic energy of the mass.

(d) Calculate the work done in pulling up the string.

(e) How is the change in mechanical energy of the mass related to the work done on it?

Mech-2

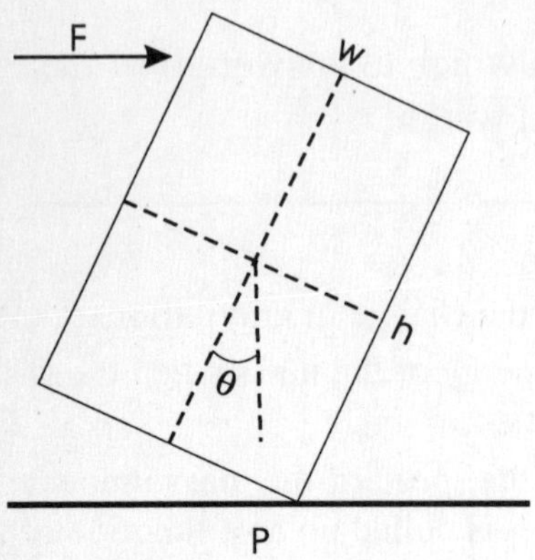

A rectangular solid of uniform density, mass *m*, and cross section *hw*, is shown in the figure above. A horizontal force *F* is applied to the solid at the corner of the solid.

(a) Find the magnitude of the torque of *F* about the pivot point *P*.

(b) Find the magnitude of the torque of the weight of the solid about *P*.

(c) Calculate the magnitude of the force *F* required to keep the mass in equilibrium as a function of the tilt angle θ.

(d) Find the angle for unstable equilibrium for the solid.

Mech-3

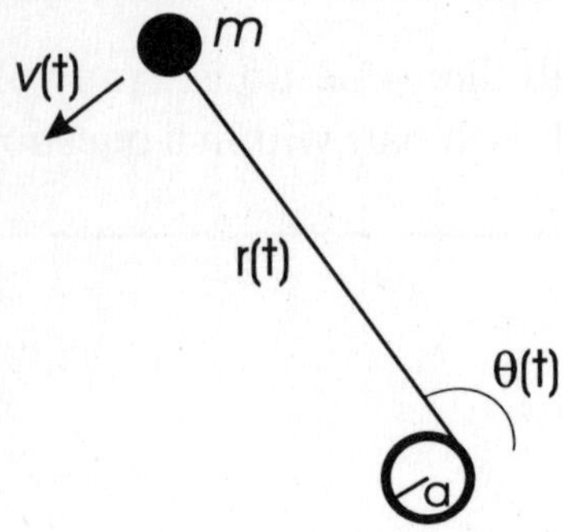

A mass *m* is at the end of a string of length *r(t)*. The other end is tethered to a post of radius *a*, as shown in the figure above. The mass has initial counterclockwise velocity v_0 and an initial radius r_0. As the mass revolves, the string is wrapped around the post and *r(t)* decreases with time.

(a) Write the angular velocity as a function of *r(t)* and constants of the motion.

(b) As the string is wrapped around the post, *r(t)* decreases. Express the time derivative of *r(t)* in terms of *r(t)* and the constants of the motion.

(c) Solve for *r(t)*, expressed in terms of the initial velocity and radius.

(d) Express the angular momentum as a function of *t*. Why is the angular momentum not conserved?

SECTION II—ELECTRICITY AND MAGNETISM

Directions: Answer all three questions. You will have 45 minutes in which to answer all of the questions. Note that each part within a question may not have equal weight.

E-1

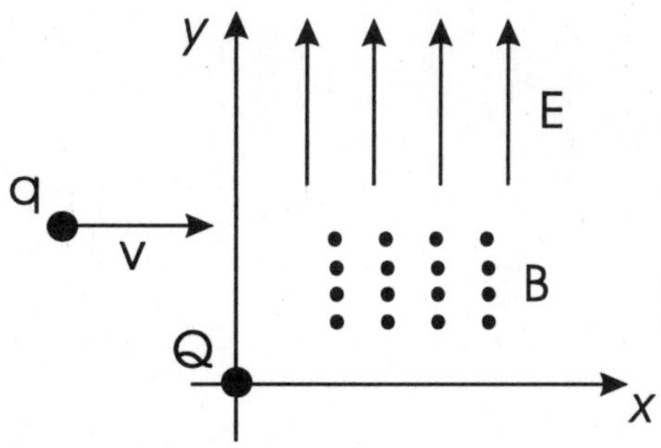

There is a uniform electric field along the *y*-direction and a uniform magnetic field pointing out of the plane of the paper, as shown in the figure above.

(a) A particle with positive charge q and mass m passes through this region with a velocity v in the $+x$-direction. For what value of v does the particle travel through the region without changing its speed or direction?

(b) If the particle travels with a lower velocity, describe the motion of the particle (qualitatively).

(c) A particle of charge Q and mass M is placed at the origin, at rest. Write the equations for the components of the force on the charge.

(d) Describe the initial motion of the charge qualitatively.

E-2

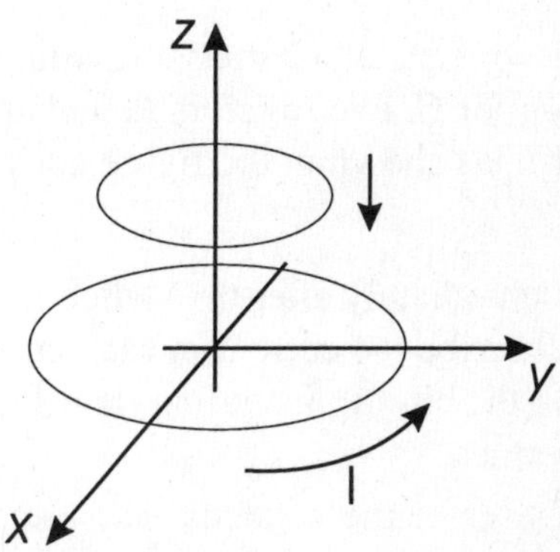

A circular, conducting loop is centered at the origin with current flowing as shown in the figure above. A smaller conducting loop is coaxial with the first, and falling toward it.

(a) What is the direction of the current induced in the smaller loop?

(b) What is the direction of the magnetic force between the loops?

(c) The small loop passes through the larger loop. What is the direction of the current when the two loops are coplanar?

(d) What is the direction of the induced current when the smaller loop is below the *x*-*y* plane?

E-3

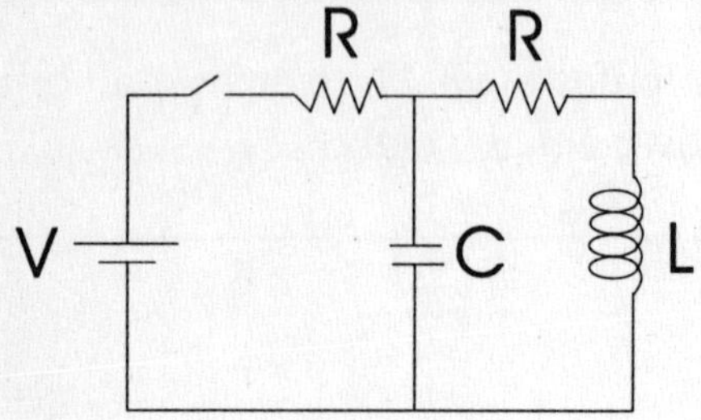

A circuit consists of a battery providing voltage V, a capacitor C, two resistors R, and an inductor L, as shown in the figure above.

(a) Immediately after the switch is closed, describe the current in each component in the circuit, including their relative values.

(b) Describe the voltages across each component when the switch is initially closed.

(c) After the circuit has reached a steady state, describe the voltages and currents in the circuit.

(d) Immediately after the switch is closed, which component has the greatest power dissipation?

Physics C Practice Test 2

ANSWERS AND EXPLANATIONS

SECTION I—MECHANICS

QUICK-SCORE ANSWERS									
1.A	8.C	15.E	22.C	29.E	36.E	43.A	50.C	57.A	64.C
2.C	9.B	16.B	23.B	30.D	37.A	44.E	51.B	58.D	65.D
3.B	10.E	17.C	24.A	31.C	38.A	45.A	52.D	59.B	66.D
4.E	11.A	18.D	25.D	32.E	39.B	46.B	53.A	60.A	67.E
5.E	12.D	19.E	26.C	33.B	40.D	47.D	54.B	61.B	68.D
6.D	13.B	20.D	27.D	34.D	41.C	48.D	55.C	62.E	69.A
7.E	14.C	21.A	28.A	35.B	42.B	49.E	56.B	63.B	70.D

1. **The correct answer is (A).** Power is the dot product of force and velocity. The power is zero because the force and velocity are perpendicular.

2. **The correct answer is (C).** As the particle is deflected by the force, the force and velocity vectors become more aligned, resulting in a positive dot product. Since the power is positive, the kinetic energy increases.

3. **The correct answer is (B).** The vertical component of velocity is initially zero. Use the equation $h = \left(\frac{1}{2}\right)gt^2$ to find the time required for the projectile to fall a distance $h = 80$m.

4. **The correct answer is (E).** The particle reverses direction when $t = \frac{v(0)}{a}$. The subsequent velocity of the particle is negative.

5. **The correct answer is (E).** The vertical velocity component is positive when Jack throws the ball. When the ball falls to the same height above the ground again, the vertical component has the same magnitude but the opposite sign. When it falls below the starting height, the velocity component is even more negative.

6. **The correct answer is (D).** The horizontal component increases as the block accelerates down the incline. The force of gravity is vertical. The only force with a horizontal component is that of the incline.

7. **The correct answer is (E).** The velocity is the derivative of the position: $v(t) = 3t^2 + 6t$. Setting this to zero and solving for t results in $t = 0$ or $t = -2$. The acceleration is the second derivative: $a(t) = 6t + 6$. Evaluating at -2 and 0 results in $a(-2) = -6$ and $a(0) = 6$.

8. **The correct answer is (C).** The net force is zero since there is no acceleration. The only forces acting on the mass are gravity and the (vertical) normal force of the surface.

9. **The correct answer is (B).** The net force on the mass is zero, so the acceleration is also zero, which implies that the velocity is constant.

10. **The correct answer is (E).** The velocity is $v(t) = 2t + 5$; the acceleration is constant, $a = 2$.

11. **The correct answer is (A).** Momentum is always conserved in the absence of external forces. Kinetic energy is only conserved in elastic collisions.

12. **The correct answer is (D).** The minimum y component of each velocity for each mass is 0. If both vectors are parallel to y, the y components are each v. So the total y component of velocity is between 0 and $2v$. Since the final momentum is zero, the y component of velocity of the third mass must be equal and opposite to the sum of the y components of the other two masses.

13. **The correct answer is (B).** The cube is in equilibrium because the net force and net torque is zero. The equilibrium is unstable because a slight perturbation results in a torque that will result in further departure from equilibrium.

14. **The correct answer is (C).** In the absence of friction, the energy required to move an object in a gravitational field only depends on the initial and final positions.

15. **The correct answer is (E).** The gravitational potential energy is converted into kinetic energy, which is then converted to elastic potential. Hence, the elastic energy is equal to gravitational energy: $mgh = \left(\frac{1}{2}\right)kx^2$, which implies that $x = \sqrt{\frac{2mgh}{k}}$.

16. **The correct answer is (B).** The gravitational potential energy is converted to kinetic energy: $\left(\frac{1}{2}\right)mv^2 = mgR$.

17. **The correct answer is (C).** Since there is no friction, the mechanical energy of the block plus spring is conserved. The block returns to its initial position.

18. **The correct answer is (D).** The center of mass of Donny and Marie remains fixed. If the origin is defined as being at the center of mass, and x is Donny's position, then

$75x = m(1 - x)$ or $m = \frac{75x}{(1-x)}$, where m is Marie's mass. The boat moves $\frac{1}{7}m$, so $x - (1 - x) = \frac{1}{7}$. Solving for x results in $x = \frac{4}{7}$. Inserting this in the previous equation results in $m = \left(\frac{4}{3}\right)75 = 100\text{kg}$.

19. **The correct answer is (E).** The acceleration is constant and equal to $\frac{10N}{5\text{kg}}$. Thus, the velocity is $2t$. The work is the time integral of the force times the velocity, $10*2t$. The kinetic energy of the block is equal to this work, 40J, added to the initial kinetic energy, 10J, which is 50J.

20. **The correct answer is (D).** The impulse is equal to the change in momentum, (0.1kg)(6m/s). Divide by the time interval, 0.01s.

21. **The correct answer is (A).** Momentum is conserved in the collision: (5kg)(2m/s) = (5kg + 15kg)v, so $v = 0.5$ m/s.

22. **The correct answer is (C).** The centripetal force is equal to the force of gravity:

$\frac{mv^2}{r} = Mg$. Solve for $v = \sqrt{\frac{Mgr}{m}} = \sqrt{16} = 4\text{m/s}$.

23. **The correct answer is (B).** The angular momentum is the moment of inertia times the angular velocity. The moment of inertia about the z-axis of each mass is $m\left(\frac{1}{2}\right)^2(\sin\theta)^2 = \left(\frac{ml^2}{4}\right)(\sin\theta)^2$. The moment of inertia is $\left(\frac{ml^2}{2}\right)(\sin\theta)^2$. The angular momentum is $ml^2\frac{w}{2}\sin^2\theta$.

24. **The correct answer is (A).** The cylindrical shell has a greater moment of inertia, hence, less acceleration.

25. **The correct answer is (D).** The rider feels weightless when the centripetal acceleration is equal to the acceleration of gravity:

$$\frac{v^2}{R} = g \text{ or } v = \sqrt{gR}$$

26. **The correct answer is (C).** The angular acceleration is the torque divided by the moment of inertia: $\frac{4\text{N}-m}{\left[0.2\text{kg}(0.5m)^2\right]} = \frac{40\text{rad}}{s^2}$. This change in the angular speed is the product of the acceleration and time: 40 rad/s^2 2s = 80 rad/s. This is added to the initial speed of 30 rad/s.

27. **The correct answer is (D).** The center of mass of the books must remain on the left of the table's edge. If the edge of the table is taken as the origin, the center mass of the books is given by $\frac{-L}{6} + \left(x - \frac{L}{2}\right) = x - \frac{2L}{3}$. The condition for equilibrium is that the location of the center of mass is less than zero: $x < \frac{2L}{3}$.

28. **The correct answer is (A).** The moment of inertia is smallest for the direction with the smallest dimension, H. The angular acceleration is inversely proportional to the moment of inertia, so it is greatest for the axis with the smallest moment.

29. **The correct answer is (E).** The mechanical energy of the oscillator remains constant.

30. **The correct answer is (D).** The periods of small oscillations in the x and y directions are independent of amplitude and equal.

31. **The correct answer is (C).** The speed of the child at the bottom of the swing can be computed by equating the gravitational potential energy at the top of the swing to the kinetic energy at the bottom: $\left(\frac{1}{2}\right)mv^2 = mgL$

or $v^2 = 2gL$

The centripetal acceleration is $\frac{v^2}{L} = 2g$.

32. **The correct answer is (E).** The distance of the satellite to the Earth's center is $2R$, where R is the radius. Since the acceleration is inversely proportional to the square of the distance, the acceleration is reduced by a factor of 4.

33. **The correct answer is (B).** The velocity of the masses after collision is determined by conservation of momentum: $Mv = 2Mv'$ or $v' = \frac{v}{2}$.

The kinetic energy after collision is reduced by $\frac{1}{2}$. The elastic potential energy at maximum extension is reduced by the same amount, resulting in an amplitude of $\frac{A}{\sqrt{2}}$.

34. **The correct answer is (D).** While kinetic and gravitational energy are conserved in circular orbits, only angular momentum is conserved for all orbits (Kepler's Second Law).

35. **The correct answer is (B).** The potential energy is half the total energy when

$$\left(\frac{1}{2}\right)kA^2 = \left(\frac{1}{2}\right)\left(\frac{1}{2}\right)kx^2 \text{ or } x = \frac{A}{\sqrt{2}}.$$

SECTION I—ELECTRICITY AND MAGNETISM

36. **The correct answer is (E).** The energy required to bring the second charge to its position is $\frac{kq^2}{d}$. The third charge requires $\frac{2kq^2}{d}$. Thus, the total is $\frac{3kq^2}{d}$.

37. **The correct answer is (A).** The net charge enclosed by a surface outside the two spheres is zero, hence, the electric field is also zero.

38. **The correct answer is (A).** The energy stored in a capacitor is proportional to the capacitance, which is proportional to K_e.

39. **The correct answer is (B).** The charge enclosed in a cylindrical surface with radius r between a and b is Q. According to Gauss's Law, the flux through the surface is given by $\frac{Q}{\varepsilon_0}$. The area of the surface is $2\pi rL$, so $2\pi rLE = \frac{Q}{\varepsilon_0}$ or $E = \frac{Q}{(2\pi rL\varepsilon_0)} = \frac{2kQ}{(rL)}$.

40. **The correct answer is (D).** The force, QE, is only along the x-direction, so the work is QEx_0.

41. **The correct answer is (C).** The electric field vector points from the positively charged plates to the negatively charged plates.

42. **The correct answer is (B).** The potential at the surface is $\frac{kQ}{R}$. The potential is constant inside the sphere. Thus, the energy required is $\frac{kqQ}{R}$.

43. **The correct answer is (A).** The potential at the origin is $\frac{kQ}{b} - \frac{kQ}{b} = 0$, so no energy is required to move a charge from infinity to the origin.

44. **The correct answer is (E).** At the midpoint, the charge q is $\frac{b}{2}$ from each of the other two charges. The two charges Q repel each other with a force $\frac{kQ^2}{b^2}$. This must be balanced by the attractive force of q:

$$\frac{kqQ}{\left(\frac{b}{2}\right)^2}+\frac{kQ^2}{b^2}=0 \text{ or } q=-\frac{1}{4}Q.$$

45. **The correct answer is (A).** Since the field is zero inside the sphere, the work required is also zero.

46. **The correct answer is (B).** Since the capacitance increases with the insertion of the dielectric, the voltage decreases.

47. **The correct answer is (D).** The work done on each mass is proportional to the charge. The mass does not affect the work done.

48. **The correct answer is (D).** The kinetic energy is transformed to potential energy. Each mass has speed $\frac{v}{2}$: $2\left(\frac{1}{2}\right)m\left(\frac{v}{2}\right)^2=\frac{kq^2}{r}$. Solving for r:

$$r=\frac{4kq^2}{\left(mv^2\right)}.$$

49. **The correct answer is (E).** The potential is constant but not zero inside the sphere, which implies that its gradient and the electric field are zero.

50. **The correct answer is (C).** An equal amount of current flows through each of the two resistors in circuit 2 and the resistor in circuit 1. Since the power is I^2R, circuit 2 dissipates twice the power that circuit 1 dissipates.

51. **The correct answer is (B).** Let r be the internal resistance of the battery. When only one resistor is connected, the current is $1\text{A} = \frac{V}{(r+R)}$. When the second resistor is connected in parallel, the external parallel resistance is $\frac{R}{2}$. Consequently, $1.5A = \frac{V}{\left(r+\frac{R}{2}\right)}$. Hence, $r+R=1.5\left(r+\frac{R}{2}\right)$ or $r=\frac{R}{2}$.

52. **The correct answer is (D).** The effective resistance of two parallel resistors R is $R_{eff}=\frac{R}{2}$. The effective resistance of two series resistors is $R_{eff}= 2R$. The power is $\frac{V^2}{R_{eff}}$. The ratio of the powers is $\frac{2V^2}{\left(\frac{V^2}{2}\right)}=4$.

53. **The correct answer is (A).** The current is initially high as the capacitor discharges, decreasing asymptotically to zero when the capacitor is fully charged.

54. **The correct answer is (B).** The charge is equally shared among the capacitors, resulting in a voltage $\frac{\left(\frac{Q}{3}\right)}{C}$.

55. **The correct answer is (C).** The capacitor blocks current flow through the resistor on the right. The two resistors in the center are in parallel, resulting in an effective resistance of $\frac{R}{2}$. This resistance is in series with R, so the total resistance is $R+\frac{R}{2}=\left(\frac{3}{2}\right)R$. According to Ohm's Law, the current is $\frac{V}{\left(\frac{3R}{2}\right)}=\left(\frac{2}{3}\right)\frac{V}{R}$.

56. **The correct answer is (B).** The resistance of the *A* leg of the circuit is 10Ω. The effective resistance of the *B* leg is 8Ω. The *C* leg has resistance $3\Omega + 6\Omega = 9\Omega$. The resistance is lowest for *B*, so the current is highest. Point *D* has only half the current of *B*, while *E* is equivalent to *A*.

57. **The correct answer is (A).** Since the voltage at the battery terminals is lower because of internal resistance, the current (hence, the power) delivered to the circuit is lower.

58. **The correct answer is (D).** The magnetic field at the charge is perpendicular to its direction of motion in a direction determined by the right-hand rule. Using the Lorentz force law, the force is found to be toward the wire.

59. **The correct answer is (B).** The flux is changing the most rapidly when the loop is parallel to the *y-z* plane. Hence, the emf is maximum.

60. **The correct answer is (A).** For points on the *z*-axis, the magnetic field of the loop is parallel to *z*. The force is on the zero since the velocity and magnetic field are parallel.

61. **The correct answer is (B).** Capacitor *A* is capable of storing more charge when it is filled with dielectric material. When the dielectric is in capacitor *B*, some of that charge flows to *B*.

62. **The correct answer is (E).** As the bar accelerates under the force of gravity, the current in the bar increases, resulting in an increasing magnetic force. Eventually, the gravitational and magnetic forces cancel, resulting in a constant velocity.

63. **The correct answer is (B).** The lower loop produces a magnetic field with a small radial outward component at the upper loop. This results in a downward force on the upper loop.

64. **The correct answer is (C).** According to the right-hand rule, the magnetic field at the charge is along $+x$. The magnetic force is given by the Lorentz force law to be in the $-z$ direction.

65. **The correct answer is (D).** The circuit's energy is constant but stored in different parts of the circuit during the oscillation. When the current is maximum, the magnetic field in the inductor is maximum, while the electric field in the capacitor is minimum.

66. **The correct answer is (D).** In the absence of currents and charges, there cannot be a static electric field. Nevertheless, there can be electromagnetic waves, so the electric and magnetic fields need not be zero. However, the existence of an electric field implies there is a magnetic field.

67. **The correct answer is (E).** The work done by a static, uniform magnetic field is always zero because the force is perpendicular to the velocity.

68. **The correct answer is (D).** The magnetic field of one wire, at the location of the other, is in one direction for points on one side of the closest point and on the other for points on the opposite side. This produces a torque.

69. **The correct answer is (A).** The current in the circuit on the left induces a current in the other inductor, which increases the effective inductance of the left-hand circuit. The increased inductance results in a lower oscillation frequency.

70. **The correct answer is (D).** Each of the two capacitors in the center stores (4μF)(10V) = 40μC. The voltage across each of the 8μF capacitors is only 5V. So, each one stores (8μF)(5V) = 40μC. The total charge stored is 40 + 40 + 40 + 40 = 160μC.

SECTION II—MECHANICS

FREE RESPONSE

Mech-1

(a) The gravitational potential changes by $mg\left(y_0 - \frac{y_0}{2}\right) = \frac{mgy_0}{2}$. The motion does not affect the potential energy.

(b) Angular momentum is conserved since the force is perpendicular to the motion. Hence, $mv_0 y_0 = mvy$ or $v = \frac{v_0 y_0}{y} = 2v_0$. The initial speed v_0 is given by:

$$v_0 = \sqrt{\frac{g}{y_0}}\theta_0 y_0 = \sqrt{gy_0}\theta_0.$$

Combine with the previous equation to find: $v = 2\sqrt{gy_0}\theta_0$

(c) The change in kinetic energy is $\frac{1}{2}m(2v_0)^2 - \frac{1}{2}mv_0^2 = \frac{3}{2}mv_0{}^2$.

(d) The required force is the sum of the gravitational and centripetal forces:

$$F = mg + m\frac{v^2}{y}.$$

Integrate from $y = y_0$ to $y = \frac{y_0}{2}$ to find the work:

$$W = -\int_{y_0}^{\frac{y_0}{2}} mg\,dy - \int_{y_0}^{\frac{y_0}{2}} \frac{mv^2}{y}dy$$

$$= \frac{1}{2}mgy_0 - mv_0^2 y_0^2 \int_{y_0}^{\frac{y_0}{2}} \frac{dy}{y^3}$$

$$= \frac{1}{2}mgy_0 + \frac{1}{2}mv_0^2 y_0^2\left[\left(\frac{1}{\frac{y_0}{2}}\right)^2 - \left(\frac{1}{y_0}\right)^2\right]$$

$$= \frac{1}{2}mgy_0 + \frac{3}{2}mv_0^2$$

(e) The change in mechanical energy is the sum of the changes in the kinetic and potential energies. The change in potential energy was computed in part (a) of this question. The change in kinetic energy was computed in part (c) of this question. Their sum is equal to the work done on the mass.

Mech-2

(a) The height of the corner above the pivot is given by $(h\cos\theta + w\sin\theta)$. Hence, the torque about P is given by $(h\cos\theta + w\sin\theta)F$.

(b) The force of gravity is vertical. The horizontal distance from the center of mass to P is given by $\left(-\frac{h}{2}\sin\theta + \frac{w}{2}\cos\theta\right)$. Hence, the torque is $\left(-\frac{h}{2}\sin\theta + \frac{w}{2}\cos\theta\right)mg$.

(c) Set these torques equal and solve for F: $F = \frac{1}{2}mg\frac{w\cos\theta - h\sin\theta}{h\cos\theta + w\sin\theta}$

(d) The solid is in equilibrium when $F = 0$. Hence, $\tan\theta = \frac{w}{h}$

Mech-3

(a) The kinetic energy is conserved since no work is done on the mass by the string. The speed of the mass remains constant, so $v_0 = r(t)\omega(t)$ and $\omega(t) = \frac{v_0}{r(t)}$.

(b) Each time the mass makes one revolution, the length of the string decreases by $2\pi a$. Hence, $\frac{dr}{dt} = -a\omega(t)$.

$$= -a\frac{v_0}{r(t)}$$

(c) Separate variables and integrate:

$$r'dr' = -av_0 dt'$$

$$\int_{r_0}^{r} r'dr' = -\int_{0}^{t} av_0 dt'$$

$$\frac{r^2}{2} - \frac{r_0^2}{2} = -av_0 t$$

$$r(t) = \sqrt{r_0^2 - 2av_0 t}$$

(d) $L(t) = mv_0 r(t)$. The angular momentum decreases with time. There is a torque on the mass about an axis through the center of the post.

SECTION II—ELECTRICITY AND MAGNETISM

E-1

(a) The electric force is in the $+y$ direction, while the magnetic force is in $-y$ direction, so they can cancel each other. According to the Lorentz force law, $F = q(E + vxB)$. Hence, if $v = \frac{E}{B}$, the electric and magnetic forces add to zero.

(b) If the velocity has a smaller magnitude, the electric force will be larger and the particle will spiral, drifting toward the $+y$ direction. You get half of the credit for y drift and half for spiral.

(c)
$$F_x = v_y B$$
$$F_y = QE - v_x B$$

(d) The initial motion of the charge is in $+y$. As it begins to move, it curves to the right. Each part is worth half credit.

E-2

(a) According to Lenz's Law, the flow of current will be opposite that of the large loop.

(b) Since the current is flowing in opposite directions, the magnetic force is repulsive.

(c) When the small loop is in the *x-y* plane, the magnetic flux is not changing, so the induced emf is zero.

(d) After passing through the *x-y* plane, the direction of the current reverses, so it is in the same direction as the current in the large loop.

E-3

(a) The initial current is greatest through the left-hand resistor, zero through the capacitor, and equal in the other resistor and inductor.

(b) The voltage across the left-hand resistor plus the voltage across the inductor and resistor equals V. The capacitor's voltage is equal to that of the right-hand resistor plus the inductor.

(c) There is equal current in the two resistors and the inductor. The current in the capacitor is zero. The voltage across the two resistors and the capacitor is equal to $\frac{V}{2}$. There is zero voltage across the inductor.

(d) Only resistors dissipate power. The current in the left-hand resistor is greater, so the power dissipation is also greater.

NOTES

NOTES

NOTES

NOTES

NOTES

NOTES